General Remarks

Lecture Notes are printed by photo-offset from the master-copy delivered in camera-ready form by the authors. For this purpose Springer-Verlag provides technical instructions for the preparation of manuscripts. See also *Editorial Policy* on the inside of the backcover.

Careful preparation of manuscripts will help keep production time short and ensure a satisfactory appearance of the finished book. The actual production of a Lecture Notes volume normally takes approximately 8 weeks.

Authors receive 50 free copies of their book. No royalty is paid on Lecture Notes volumes.

Authors are entitled to purchase further copies of their book and other Springer mathematics books for their personal use, at a discount of 33,3 % directly from Springer-Verlag.

Commitment to publish is made by letter of intent rather than by signing a formal contract. Springer-Verlag secures the copyright for each volume.

Addresses:

Professor M. Griebel
Institut für Angewandte Mathematik
der Universität Bonn
Wegelerstr. 6
D-53115 Bonn, Germany
e-mail: griebel@iam.uni-bonn.de

Professor D. E. Keyes
Computer Science Department
Old Dominion University
Norfolk, VA 23529–0162, USA
e-mail: keyes@cs.odu.edu

Professor R. M. Nieminen
Laboratory of Physics
Helsinki University of Technology
02150 Espoo, Finland
e-mail: rniemine@csc.fi

Professor D. Roose
Department of Computer Science
Katholieke Universiteit Leuven
Celestijnenlaan 200A
3001 Leuven-Heverlee, Belgium
e-mail: dirk.roose@cs.kuleuven.ac.be

Professor T. Schlick
Department of Chemistry and
Courant Institute of Mathematical Sciences
New York University
and Howard Hughes Medical Institute
251 Mercer Street, Rm 509
New York, NY 10012-1548, USA
e-mail: schlick@nyu.edu

Springer-Verlag, Mathematics Editorial
Tiergartenstrasse 17
D-69121 Heidelberg, Germany
Tel.: *49 (6221) 487-185
e-mail: peters@springer.de
http://www.springer.de/math/peters.html

Lecture Notes in Computational Science and Engineering

1

Editors
M. Griebel, Bonn
D. E. Keyes, Norfolk
R. M. Nieminen, Espoo
D. Roose, Leuven
T. Schlick, New York

Springer
Berlin
Heidelberg
New York
Barcelona
Budapest
Hong Kong
London
Milan
Paris
Santa Clara
Singapore
Tokyo

Daniele Funaro

Spectral Elements for Transport-Dominated Equations

With 97 Figures and 12 Tables

 Springer

Author

Daniele Funaro
Dipartimento di Matematica
Università di Modena
Via Campi 213/B
41100 Modena, Italy
e-mail: funaro@unimo.it
http://www.matematica.unimo.it/pfunaeng.htm

Cataloging-in-Publication Data applied for

Die Deutsche Bibliothek - CIP-Einheitsaufnahme
Funaro, Daniele:
Spectral elements for transport dominated equations / Daniele
Funaro. - Berlin ; Heidelberg ; New York ; Barcelona ;
Budapest ; Hong Kong ; London ; Milan ; Paris ; Santa Clara ;
Singapore ; Tokyo : Springer, 1997
  (Lecture notes in computational science and engineering ; 1)
  ISBN 3-540-62649-2
NE: GT

Front cover photo by Micol Pennacchio

Mathematics Subject Classification (1991): primary: 65M70, 65N22
                                          secondary: 76D10, 65D99, 65Y99

ISBN 3-540-62649-2 Springer-Verlag Berlin Heidelberg New York

Cover Design: Friedhelm Steinen-Broo, Estudio Calamar, Spain
Cover: *design & production* GmbH, Heidelberg
Typesetting: Camera-ready by the author

SPIN 10565167   41/3143 – 5 4 3 2 1 0 – Printed on acid-free paper

Preface

In the last few years there has been a growing interest in the development of numerical techniques appropriate for the approximation of differential model problems presenting *multiscale* solutions. This is the case, for instance, with functions displaying a smooth behavior, except in certain regions where sudden and sharp variations are localized. Typical examples are *internal* or *boundary layers*. When the number of degrees of freedom in the discretization process is not sufficient to ensure a fine resolution of the layers, some *stabilization* procedures are needed to avoid unpleasant oscillatory effects, without adding too much *artificial viscosity* to the scheme. In the field of finite elements, the *streamline diffusion* method, the *Galerkin least-squares* method, the *bubble function* approach, and other recent similar techniques provide excellent treatments of transport equations of elliptic type with small diffusive terms, referred to in fluid dynamics as *advection-diffusion* (or *convection-diffusion*) equations.

Goals

This book is an attempt to guide the reader in the construction of a computational code based on the spectral collocation method, using algebraic polynomials. The main topic is the approximation of elliptic type boundary-value partial differential equations in 2-D, with special attention to transport-diffusion equations, where the second-order diffusive terms are strongly dominated by the first-order advective terms. Applications will be considered especially in the case where nonlinear systems of partial differential equations can be reduced to a sequence of transport-diffusion equations. Examples of simple and more advanced problems, drawn from various fields of computational physics and engineering, are scattered throughout the text. We examine both the implementation on simple domains and the extension to more complicated geometries through the *domain decomposition* approach. The main point is the introduction of suitable *upwind* collocation grids associated with the nonsymmetric differential operator. For the Laplace operator, the method simplifies to the ordinary version of the spectral elements for conforming decompositions, for which suitable iterative algorithms are suggested. Thus, the proposed algorithms are not alternatives to the usual methods, but rather generalizations of them, offering the possibility of obtaining low-cost solutions when other non stabilized techniques fail. Much theory has yet to be done, so the subject may have attractions for mathematicians involved in theoretical work, while our treatment is more heuristic.

This book is focused on the specific subject of Legendre collocation approximations and contains many unpublished results. Although it is not comprehensive with the respect to the entire range of spectral element computations, it can also be employed as an introduction to this field. Most of the basic ingredients of the Legendre collocation method are discussed, as well as a certain number of new ideas which the reader can expand or adapt to his or her own code.

Organization

The chapters are organized as follows. In the first chapter, we show how to approximate the Poisson problem, defined in $\Omega =]-1,1[\times]-1,1[$ with Dirichlet boundary conditions, by the collocation method at the Legendre nodes. We recall some known theoretical results and we explain in detail how to efficiently recover, by a preconditioned iterative procedure, the approximating polynomials. The adaptation of the results for the Poisson equation to the case of steady transport-diffusion equations is accomplished in chapter two. The aim is to find meaningful approximate solutions both near and far from the *boundary layers*, even if the degrees of freedom are not enough for an accurate resolution, due to the heavy dominance of transport terms. We succeed in this by collocating the equation at a special set of points, called the *upwind grid*, obtained by deforming the Legendre nodes in Ω according to the size of the coefficients of the differential operator, taking care of the magnitude and the direction of the *flux* indicated by the transport terms. The upwind grid also plays a crucial role in the construction of a preconditioning matrix, allowing for a fast convergence in the iterative solution of the linear system corresponding to the collocation problem. We deal with the treatment of Neumann and other mixed-type boundary conditions in chapter three.

For more complicated domains Ω, we use the spectral element method based on a conforming decomposition in quadrilateral subdomains which, mapped to a reference square, bring us back to the implementation of the collocation method. As described in chapter four, the various pieces of the solution are joined together by weakly imposing the continuity of the global approximating function and its derivatives on the _interfaces_. The numerical determination of the whole solution is carried out by an iterative algorithm, decoupling the global system into a sequence of collocation problems defined in each subdomain. A speed-up of the convergence is realized by suitable preconditioners applied to the interface unknowns. Special grids at the interfaces are also introduced to handle the case of transport-dominated equations.

Some classical examples of nonlinear time-dependent equations, such as Navier-Stokes, are examined in chapter five. We will be concerned with the approximation of both the evolutive and the steady state solutions.

Some more experiments, observations, generalizations, and other related questions are collected in chapter six, while in the appendix the reader finds a list of the properties characterizing the family of Legendre polynomials used throughout the text.

A very basic perspective on spectral methods

In order to show that employing spectral methods is not as complicated as skeptics may claim, we briefly introduce in a short and simple way the principles inspiring these approximation techniques. We report here a series of elementary results.

Denoting by $\mathbf{P}_n$, $n \geq 1$, the space of polynomials in $[-1, 1]$ of degree not greater than n, the following fundamental properties hold:

THEOREM 1 - _The dimension of $\mathbf{P}_n$ is $n + 1$._

THEOREM 2 - _The derivative of $q_n \in \mathbf{P}_n$ is a polynomial $\frac{d}{dx}q_n$ belonging to $\mathbf{P}_n$ (in particular $\frac{d}{dx}q_n \in \mathbf{P}_{n-1}$, but this is not a crucial point!)._

THEOREM 3 - _The derivative operator $\frac{d}{dx}$ is a linear application from $\mathbf{P}_n$ into $\mathbf{P}_n$. Hence, for a given basis in $\mathbf{P}_n$, there exists a $(n + 1) \times (n + 1)$ matrix representing $\frac{d}{dx}$._

THEOREM 4 - _The k^{th} derivative operator $\frac{d^k}{dx^k}$ is represented in $\mathbf{P}_n$ by the kth power of the first derivative matrix._

In other words, the last two statements say that whenever we need to evaluate the <u>exact</u> derivatives of $q_n \in \mathbf{P}_n$, expressed in a certain basis, we simply apply our matrix a certain number of times to get $\frac{d^k}{dx^k}q_n$. The choice of a basis in $\mathbf{P}_n$ decides which technique we are going to employ. For instance, the collocation method is obtained by fixing $n + 1$ nodes in $[-1, 1]$ and taking the canonical Lagrange polynomial basis with respect to this set of points (i.e., the jth polynomial takes the value 1 at the jth node and 0 at the remaining nodes). For several reasons, mainly suggested by well-known and powerful results of approximation theory, it is convenient to choose the nodes associated with the zeroes of orthogonal families of polynomials, such

as the Chebyshev or the Legendre polynomials. The numerical approximation of a simple linear differential problem is simply the technical matter of
evaluating the differentiation matrix in $\mathbf{P}_n$, setting up the differential operator, imposing, in some convenient way, the boundary conditions, and solving
the final linear system. Some software is available, for instance, in FUNARO
(1993b) and some additional is provided by W.–S. DON and A. SOLOMONOFF
at http://www.cfm.brown.edu/people/wsdon/pseudopack_v2.1.html.

Of course, many steps remain between these simple ideas and the practical
codes. This is what we wish to explain in the following pages.

Acknowledgments

The author wishes to thank Roger Peyret, Paola Pietra and Saul Abarbanel
for consulting on various topics. Additional thanks go to Michèle Mulrooney
and David Keyes for assistance in proofreading the manuscript.

Modena, December 1996

Contents

The Poisson Equation in the Square

We begin by examining the numerical approximation, using the Legendre collocation method, of a simple boundary-value problem defined in a square. This analysis is the starting point for several generalizations to be developed in the following chapters.

1.1 Statement of the problem and preliminary results

This chapter concerns the partial differential equation

$$(1.1.1) \qquad -\Delta U = f \qquad \text{in} \quad \Omega,$$

where $\Delta := \frac{\partial^2}{\partial x^2} + \frac{\partial^2}{\partial y^2}$ is the *Laplace operator* and Ω is the square $]-1, 1[\times]-1, 1[$. We provide problem (1.1.1) with the non-homogeneous *Dirichlet* boundary conditions

$$(1.1.2) \qquad \begin{cases} U(x, -1) = g_1(x) & \forall x \in [-1, 1[, \\[2mm] U(1, y) = g_2(y) & \forall y \in [-1, 1[, \\[2mm] U(x, 1) = g_3(x) & \forall x \in]-1, 1], \\[2mm] U(-1, y) = g_4(y) & \forall y \in]-1, 1]. \end{cases}$$

In the above equations, $f : \Omega \to \mathbf{R}$ and $g_k : [-1, 1] \to \mathbf{R}$, $1 \leq k \leq 4$, are suitable given functions.

Concerning the existence of the solution $U : \bar{\Omega} \to \mathbf{R}$ of (1.1.1)-(1.1.2) we mention a classical result, the proof of which is given in WEINBERGER (1965).

THEOREM 1.1.1 - *Let f be a continuous function in Ω satisfying $\int_\Omega f^2 dx dy$ $< +\infty$, and let g_k, $1 \le k \le 4$, be continuous functions in $[-1,1]$ with $g_1(1) = g_2(-1)$, $g_2(1) = g_3(1)$, $g_3(-1) = g_4(1)$, $g_4(-1) = g_1(-1)$, then there exists a unique solution $U \in C^0(\bar{\Omega}) \cap C^1(\Omega)$ of (1.1.1)-(1.1.2).*

Another approach to studing the theory and applications related to the *Poisson equation* (1.1.1)-(1.1.2) is the variational method. Since this subject is extensively discussed in numerous publications (for example LIONS and MAGENES (1972)), we only recall some basic ideas. The analysis is based on a main general result, attributed to P.D. Lax and N. Milgram, which can be summarized in the following statement.

THEOREM 1.1.2 - *Let us assume that $\mathbf{X}$ is a Hilbert space. Let $B : \mathbf{X} \times \mathbf{X} \to \mathbf{R}$ be a bilinear form and $F : \mathbf{X} \to \mathbf{R}$ be a linear operator. If there exist three positive constants C_1, C_2, C_3 such that*

$$(1.1.3) \quad \begin{cases} |B(U,\phi)| \le C_1 \, \|U\|_{\mathbf{X}} \, \|\phi\|_{\mathbf{X}} & \forall U \in \mathbf{X}, \, \forall \phi \in \mathbf{X}, \\[2mm] B(U,U) \ge C_2 \, \|U\|_{\mathbf{X}}^2 & \forall U \in \mathbf{X}, \\[2mm] |F(\phi)| \le C_3 \, \|\phi\|_{\mathbf{X}} & \forall \phi \in \mathbf{X}, \end{cases}$$

then there exists a unique solution $U \in \mathbf{X}$ of the equation

$$(1.1.4) \qquad\qquad B(U,\phi) = F(\phi) \qquad \forall \phi \in \mathbf{X}.$$

In addition, there exists a positive constant $C_4 > 0$ such that

$$(1.1.5) \qquad\qquad \|U\|_{\mathbf{X}} \le C_4 \sup_{\substack{\phi \in \mathbf{X} \\ \phi \neq 0}} \frac{|F(\phi)|}{\|\phi\|_{\mathbf{X}}}.$$

To see the analogy with the Poisson equation, we note that the Green formulas enable us to write

$$(1.1.6) \quad B(U,\phi) := \int_\Omega \left(\frac{\partial U}{\partial x} \frac{\partial \phi}{\partial x} + \frac{\partial U}{\partial y} \frac{\partial \phi}{\partial y} \right) dx dy = \int_\Omega f\phi \, dx dy =: F(\phi),$$

for any $\phi \in C^0(\bar{\Omega}) \cap C^1(\Omega)$ with $\phi \equiv 0$ on the boundary of Ω.

Therefore, U is expected to be the solution of a problem like (1.1.4) belonging to a certain functional space $\mathbf{X}$, in such a way that the conditions in (1.1.3) are satisfied. If, for simplicity, we assume that $g_k = 0$, $1 \leq k \leq 4$, then the correct space for the solution U and the *test functions* ϕ is the *Sobolev space* $\mathbf{X} \equiv H^1_0(\Omega)$. A comprehensive description of the Poisson equation from the derivation of the physics model to its theoretical analysis is given in DAUTRAY and LIONS (1988-1993).

In addition, always assuming that $g_k = 0$, $1 \leq k \leq 4$, by taking $\phi = U$ in (1.1.6), one recovers the *a priori* estimate

$$(1.1.7) \qquad \int_\Omega |\vec{\nabla} U|^2 \, dxdy \leq C_5 \int_\Omega f^2 \, dxdy,$$

where C_5 is a positive constant and $\vec{\nabla} U \equiv \left(\frac{\partial}{\partial x} U, \frac{\partial}{\partial y} U\right)$. Actually, (1.1.7) is a byproduct of the Schwarz inequality and the following *Poincaré* inequality

$$(1.1.8) \qquad \int_\Omega \phi^2 \, dxdy \leq C_5 \int_\Omega |\vec{\nabla}\phi|^2 \, dxdy,$$

which holds for functions ϕ vanishing on $\partial\Omega$. Another way to arrive at (1.1.7) is to use (1.1.5). We will come back later to the formulation (1.1.6) in section 1.3.

1.2 The collocation method for the Poisson equation

We describe how to apply the *collocation method* to find an approximated solution to problem (1.1.1)-(1.1.2). Several other spectral type techniques are available for the numerical treatment of the Poisson equation and the problems that will be considered later in this book. Some of these techniques can also perform better than the one examined here. However, our goal is not to present a comprehensive treatise on spectral methods (such a project has already been attempted by other authors) but rather to confine ourselves to the use of a single methodology and develop some new tools at a uniform level. With the aim of giving an introductive overview, the book of BOYD (1989), the paper of FORNBERG and SLOAN (1994), and the book of FORNBERG (1996) show, with numerous examples and comparisons, how to apply spectral methods with different approaches. These publications are therefore more suited for the novice. The reader interested in additional spectral-type approximation techniques for elliptic problems, with details of convergence theorems, may consult the books of CANUTO, HUSSAINI, QUARTERONI and ZANG (1988), BERNARDI and MADAY (1992), and FUNARO (1992).

Throughout this book, we retain the notation introduced in FUNARO (1992). A list of symbols is provided at page 194. In the appendix the reader finds a short review of the properties concerning the family of Legendre polynomials P_n, $n \in \mathbf{N}$.

Let $n \geq 1$ be an integer, then we define $\mathbf{P}_n$ to be the space of polynomials in $[-1,1]$ of degree less than or equal to n, and $\mathbf{P}_n^\star$ to be the space of polynomials in $\bar{\Omega}$ of degree less than or equal to n in each variable. The dimension of $\mathbf{P}_n^\star$ is $(n+1)^2$. We recall that the derivative of the Legendre polynomial $P_n' \in \mathbf{P}_{n-1}$ has $n-1$ distinct real roots in $]-1,1[$, which will be ordered increasingly and denoted by $\eta_j^{(n)}$, $1 \leq j \leq n-1$. Finally, one defines $\eta_0^{(n)} := -1$ and $\eta_n^{(n)} := 1$ (see also section $A.2$). From now on, the set of points $(\eta_j^{(n)}, \eta_i^{(n)}) \in \bar{\Omega}$, $0 \leq i \leq n$, $0 \leq j \leq n$, will be referred to as the *Legendre grid*. The points inside Ω are $(n-1)^2$, while the boundary points are $4n$. We give in Fig. 1.2.1 the distribution of the Legendre grid points of $\bar{\Omega}$ for $n = 10$.

We *collocate* the equations (1.1.1)-(1.1.2) at the Legendre grid, which means that we are looking for a polynomial $q_n \in \mathbf{P}_n^\star$ satisfying the set of $(n-1)^2$ linear equations

$$(1.2.1) \quad (-\Delta q_n)(\eta_j^{(n)}, \eta_i^{(n)}) = f(\eta_j^{(n)}, \eta_i^{(n)}) \quad 1 \leq i \leq n-1, \ 1 \leq j \leq n-1,$$

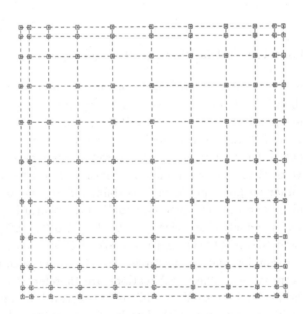

FIG. 1.2.1 - *The Legendre grid for* $n = 10$.

together with the $4n$ boundary relations

$$(1.2.2) \quad \begin{cases} q_n(\eta_j^{(n)}, \eta_0^{(n)}) = g_1(\eta_j^{(n)}) & 0 \le j \le n-1, \\[2mm] q_n(\eta_n^{(n)}, \eta_i^{(n)}) = g_2(\eta_i^{(n)}) & 0 \le i \le n-1, \\[2mm] q_n(\eta_j^{(n)}, \eta_n^{(n)}) = g_3(\eta_j^{(n)}) & 1 \le j \le n, \\[2mm] q_n(\eta_0^{(n)}, \eta_i^{(n)}) = g_4(\eta_i^{(n)}) & 1 \le i \le n. \end{cases}$$

We will check in section 1.3 that (1.2.1)-(1.2.2) actually admits a unique polynomial solution and that q_n is an approximation of U.

The collocation points may be ordered as follows

$$(1.2.3) \quad \Theta_k^{(n)} \equiv (\eta_j^{(n)}, \eta_i^{(n)}) \quad 0 \le k \le n_T := (n+1)^2 - 1 \text{ with } k = (n+1)i + j.$$

Of course, it is sufficient to compute the unknown polynomial q_n at the points $\Theta_k^{(n)}$, $0 \le k \le n_T$, and by interpolation obtain the values in other points. To do this we use formula $(A.3.3)$.

At this point we would like to show that (1.2.1)-(1.2.2) can be written as a linear system of dimension $(n+1)^2$, where the unknown vector is $\vec{X}_n :=$ $\{q_n(\Theta_k^{(n)})\}_{0 \le k \le n_T}$. We start by noting that for $0 \le i \le n$, $0 \le j \le n$ one has

$$(1.2.4) \quad -(\Delta q_n)(\eta_j^{(n)}, \eta_i^{(n)}) = -\sum_{m=0}^{n} [\tilde{d}_{jm}^{(2)} q_n(\eta_m^{(n)}, \eta_i^{(n)}) + \tilde{d}_{im}^{(2)} q_n(\eta_j^{(n)}, \eta_m^{(n)})],$$

where, according to section $A.3$ of the appendix, the coefficients $\tilde{d}_{lm}^{(2)}$, $0 \le l \le n$, $0 \le m \le n$, are the entries of the matrix $\tilde{D}_n^2$, representing the second derivative operator in the space of polynomials $\mathbf{P}_n$. Such a matrix allows the exact computation of the second derivative of a polynomial $r \in \mathbf{P}_n$ at the points $\eta_l^{(n)}$, $0 \le l \le n$, starting from the vector of the values: $r(\eta_m^{(n)})$, $0 \le m \le n$. The entries of $\tilde{D}_n^2$ can be computed and stored once and for all. The operator $\tilde{D}_n$ is the *differentiation matrix* mentioned in the preface.

Using (1.2.4) we can easily define the entries of the matrix, denoted by $\mathcal{L}_n$, corresponding to the Laplace operator in the space $\mathbf{P}_n^*$. We soon realize that $\mathcal{L}_n$ has a structure of the type presented in Fig. 1.2.2 for $n = 4$, where the dots $\bullet$ denote the nonzero entries.

We recall that the *tensor product* of two $(n+1) \times (n+1)$ matrices $A = \{a_{ij}\}_{\substack{0 \le i \le n \\ 0 \le j \le n}}$ and $B = \{b_{ij}\}_{\substack{0 \le i \le n \\ 0 \le j \le n}}$ is the $(n+1)^2 \times (n+1)^2$ matrix expressed by blocks as:

$$(1.2.5) \qquad A \otimes B := \begin{bmatrix} a_{00}B & a_{01}B & \cdots & a_{0n}B \\ a_{10}B & a_{11}B & \cdots & a_{1n}B \\ \vdots & \vdots & & \vdots \\ a_{n0}B & a_{n1}B & \cdots & a_{nn}B \end{bmatrix}$$

Then, $\mathcal{L}_n$ turns out to be the tensor product $-(I_n \otimes \tilde{D}_n^2 + \tilde{D}_n^2 \otimes I_n)$, where I_n is the $(n+1) \times (n+1)$ identity matrix. In order to impose the boundary conditions in (1.2.2), we set to zero all the rows of $\mathcal{L}_n$ corresponding to the boundary nodes, with the exception of the entries belonging to the diagonal which are set to 1 (see also section 1.7). We end up with a non-symmetric matrix, denoted by $\mathcal{A}_n$, which for $n = 4$ has the structure of Fig. 1.2.3.

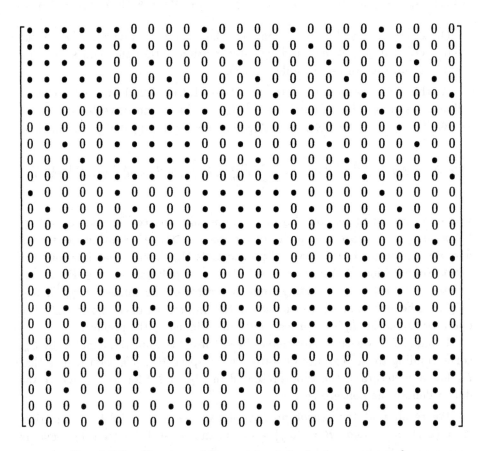

FIG. 1.2.2 - *Structure of the matrix of the Laplace operator* $\mathcal{L}_4$.

Finally, we have to specify the right-hand side vector, which will be denoted by $\vec{B}_n$. The entries of $\vec{B}_n$ take the values of the functions f, g_1, g_2, g_3, g_4 at the collocation nodes, depending on whether these nodes are inside or at the boundary $\partial\Omega$. We show for instance the right-hand side vector in the case $n = 4$, i.e.

$$
\begin{aligned}
\vec{B}_4 \equiv \big(& g_1(\eta_0^{(n)}),\ g_1(\eta_1^{(n)}),\ g_1(\eta_2^{(n)}),\ g_1(\eta_3^{(n)}),\ g_2(\eta_0^{(n)}), \\
& g_4(\eta_1^{(n)}),\ f(\Theta_6^{(n)}),\ f(\Theta_7^{(n)}),\ f(\Theta_8^{(n)}),\ g_2(\eta_1^{(n)}), \\
& g_4(\eta_2^{(n)}),\ f(\Theta_{11}^{(n)}),\ f(\Theta_{12}^{(n)}),\ f(\Theta_{13}^{(n)}),\ g_2(\eta_2^{(n)}), \\
& g_4(\eta_3^{(n)}),\ f(\Theta_{16}^{(n)}),\ f(\Theta_{17}^{(n)}),\ f(\Theta_{18}^{(n)}),\ g_2(\eta_3^{(n)}), \\
& g_4(\eta_4^{(n)}),\ g_3(\eta_1^{(n)}),\ g_3(\eta_2^{(n)}),\ g_3(\eta_3^{(n)}),\ g_3(\eta_4^{(n)}) \big).
\end{aligned}
$$

```
⎡ 1 0 0 0 0 0 0 0 0 0 0 0 0 0 0 0 0 0 0 0 0 0 0 0 0 ⎤
⎢ 0 1 0 0 0 0 0 0 0 0 0 0 0 0 0 0 0 0 0 0 0 0 0 0 0 ⎥
⎢ 0 0 1 0 0 0 0 0 0 0 0 0 0 0 0 0 0 0 0 0 0 0 0 0 0 ⎥
⎢ 0 0 0 1 0 0 0 0 0 0 0 0 0 0 0 0 0 0 0 0 0 0 0 0 0 ⎥
⎢ 0 0 0 0 1 0 0 0 0 0 0 0 0 0 0 0 0 0 0 0 0 0 0 0 0 ⎥
⎢ 0 0 0 0 0 1 0 0 0 0 0 0 0 0 0 0 0 0 0 0 0 0 0 0 0 ⎥
⎢ 0 • 0 0 0 • • • • • 0 • 0 0 0 0 • 0 0 0 0 • 0 0 0 ⎥
⎢ 0 0 • 0 0 • • • • • 0 0 • 0 0 0 0 • 0 0 0 0 • 0 0 ⎥
⎢ 0 0 0 • 0 • • • • • 0 0 0 • 0 0 0 0 • 0 0 0 0 • 0 ⎥
⎢ 0 0 0 0 0 0 0 0 0 1 0 0 0 0 0 0 0 0 0 0 0 0 0 0 0 ⎥
⎢ 0 0 0 0 0 0 0 0 0 0 1 0 0 0 0 0 0 0 0 0 0 0 0 0 0 ⎥
⎢ 0 • 0 0 0 0 • 0 0 0 • • • • • 0 • 0 0 0 0 • 0 0 0 ⎥
⎢ 0 0 • 0 0 0 0 • 0 0 • • • • • 0 0 • 0 0 0 0 • 0 0 ⎥
⎢ 0 0 0 • 0 0 0 0 • 0 • • • • • 0 0 0 • 0 0 0 0 • 0 ⎥
⎢ 0 0 0 0 0 0 0 0 0 0 0 0 0 0 1 0 0 0 0 0 0 0 0 0 0 ⎥
⎢ 0 0 0 0 0 0 0 0 0 0 0 0 0 0 0 1 0 0 0 0 0 0 0 0 0 ⎥
⎢ 0 • 0 0 0 0 • 0 0 0 0 • 0 0 0 • • • • • 0 • 0 0 0 ⎥
⎢ 0 0 • 0 0 0 0 • 0 0 0 0 • 0 0 • • • • • 0 0 • 0 0 ⎥
⎢ 0 0 0 • 0 0 0 0 • 0 0 0 0 • 0 • • • • • 0 0 0 • 0 ⎥
⎢ 0 0 0 0 0 0 0 0 0 0 0 0 0 0 0 0 0 0 0 1 0 0 0 0 0 ⎥
⎢ 0 0 0 0 0 0 0 0 0 0 0 0 0 0 0 0 0 0 0 0 1 0 0 0 0 ⎥
⎢ 0 0 0 0 0 0 0 0 0 0 0 0 0 0 0 0 0 0 0 0 0 1 0 0 0 ⎥
⎢ 0 0 0 0 0 0 0 0 0 0 0 0 0 0 0 0 0 0 0 0 0 0 1 0 0 ⎥
⎢ 0 0 0 0 0 0 0 0 0 0 0 0 0 0 0 0 0 0 0 0 0 0 0 1 0 ⎥
⎣ 0 0 0 0 0 0 0 0 0 0 0 0 0 0 0 0 0 0 0 0 0 0 0 0 1 ⎦
```

FIG. 1.2.3 - *Structure of the matrix $\mathcal{A}_4$ of the Laplace operator with boundary conditions.*

Thus, the collocation method amounts to finding the solution of the linear system

$$(1.2.6) \qquad \mathcal{A}_n \vec{X}_n = \vec{B}_n, \qquad n \geq 1.$$

An algorithm will be developed in section 1.4 for the numerical treatment of (1.2.6).

1.3   Convergence analysis of the collocation method

We define $\mathbf{P}_n^{*,0}$ to be the subset of $\mathbf{P}_n^*$ consisting of those polynomials vanishing at the boundary $\partial\Omega$. Note that a polynomial belongs to $\mathbf{P}_n^{*,0}$ if and only if it vanishes at the boundary nodes.

We begin by proving that the system (1.2.6) has a unique solution. Actually, we show a stronger result:

THEOREM 1.3.1 - *For any $n \geq 1$, the matrix $\mathcal{A}_n$ has real and strictly positive eigenvalues.*

Proof - It is clear that 1 is an eigenvalue of $\mathcal{A}_n$ of multiplicity $4n$. In this case, the eigenvectors have all the entries equal to zero except for one. The other $(n-1)^2$ eigenvalues are related to polynomials vanishing at all the boundary nodes. Therefore, we look for $\psi_n \in \mathbf{P}_n^{*,0}$, $\psi_n \neq 0$, satisfying the eigenvalue problem

$$(1.3.1) \qquad (-\Delta \psi_n)(\eta_j^{(n)}, \eta_i^{(n)}) = \lambda \psi_n(\eta_j^{(n)}, \eta_i^{(n)}),$$

for $1 \leq i \leq n-1$, $1 \leq j \leq n-1$. We observe that ψ_n is complex. Then, we use the quadrature formula $(A.2.3)$ found in the appendix. We multiply both terms of (1.3.1) by $\bar{\psi}_n(\eta_j^{(n)}, \eta_i^{(n)})\tilde{w}_j^{(n)}\tilde{w}_i^{(n)}$, where $\bar{\psi}_n$ denotes the complex conjugate of ψ_n. Summing up and recalling that $\bar{\psi}_n$ vanishes at the boundary nodes, we get

$$(1.3.2) \qquad -\sum_{i,j=1}^{n-1}(\Delta\psi_n\,\bar{\psi}_n)(\eta_j^{(n)}, \eta_i^{(n)})\,\tilde{w}_j^{(n)}\tilde{w}_i^{(n)}$$

$$= -\sum_{i,j=0}^{n}(\Delta\psi_n\,\bar{\psi}_n)(\eta_j^{(n)}, \eta_i^{(n)})\,\tilde{w}_j^{(n)}\tilde{w}_i^{(n)} =$$

$$= - \sum_{i=0}^{n} \left[\int_{-1}^{1} \left(\frac{\partial^2 \psi_n}{\partial x^2} \ \bar{\psi}_n \right) (x, \eta_i^{(n)}) dx \right] \tilde{w}_i^{(n)}$$

$$- \sum_{j=0}^{n} \left[\int_{-1}^{1} \left(\frac{\partial^2 \psi_n}{\partial y^2} \ \bar{\psi}_n \right) (\eta_j^{(n)}, y) dy \right] \tilde{w}_j^{(n)}$$

$$= \sum_{i=0}^{n} \left[\int_{-1}^{1} \left| \frac{\partial \psi_n}{\partial x} \right|^2 (x, \eta_i^{(n)}) dx \right] \tilde{w}_i^{(n)} + \sum_{j=0}^{n} \left[\int_{-1}^{1} \left| \frac{\partial \psi_n}{\partial y} \right|^2 (\eta_j^{(n)}, y) dy \right] \tilde{w}_j^{(n)}$$

$$= \lambda \sum_{i,j=1}^{n-1} |\psi_n|^2 (\eta_j^{(n)}, \eta_i^{(n)}) \ \tilde{w}_j^{(n)} \tilde{w}_i^{(n)},$$

where we used the fact that $\left[\left(\frac{\partial^2}{\partial x^2} \psi_n \right) \bar{\psi}_n \right] (x, \eta_i^{(n)})$, $0 \le i \le n$, $x \in [-1, 1]$, as well as $\left[\left(\frac{\partial^2}{\partial y^2} \psi_n \right) \bar{\psi}_n \right] (\eta_j^{(n)}, y)$, $0 \le j \le n$, $y \in [-1, 1]$, are complex polynomials of degree less than or equal to $2n - 2$. In this way, we passed from the quadratures to the integrals by successively integrating by parts. Examining the last two terms of (1.3.2), we realize that λ must be real and positive.

As a corollary, we know that the set of equations (1.2.1)-(1.2.2) has a unique solution $q_n \in \mathbf{P}_n^\star$. We can decompose q_n into the sum of two polynomials $q_n = r_n + s_n$, where $s_n \in \mathbf{P}_n^{\star,0}$ satisfies $-\Delta s_n = f$ at the nodes inside Ω, and r_n has the same boundary conditions of q_n and satisfies $-\Delta r_n = 0$ at the nodes inside Ω.

Assuming homogeneous boundary conditions, i.e. $g_k = 0$, $1 \le k \le 4$ (hence $r_n = 0$), we now briefly describe how to prove that $q_n = s_n$ converges to U for $n \to +\infty$. First, we need to introduce the *interpolant* $\tilde{I}_n f \in \mathbf{P}_n^\star$ of f at the nodes $\Theta_k^{(n)}$, $0 \le k \le n_T$. If f is continuous, one can show that

$$(1.3.3) \qquad \int_\Omega (\tilde{I}_n f - f)^2 dx dy \ \to \ 0 \qquad \text{for } n \to +\infty.$$

The more regular the function f is, the faster the error (1.3.3) decays. The rate of convergence is $n^{-\sigma}$ (or faster), where the parameter $\sigma > 0$ has an upper bound depending only on the regularity of f. If f is analytic, then the decay is of exponential type. Such behavior is typical of spectral methods and is referred to as *spectral convergence rate*. An estimate of the error between f and its interpolant is presented for the one-dimensional case in section A.4. Estimates in the two-dimensional case are provided in CANUTO and QUARTERONI (1982), MADAY (1991), BERNARDI and MADAY (1992).

We are ready to give a *stability* result, i.e. that a certain norm of s_n is bounded by a quantity that does not depend on n (note the analogy with (1.1.7)).

THEOREM 1.3.2 - *There exists a constant $C > 0$ such that, for any $n \geq 1$, the polynomial $s_n \in \mathbf{P}_n^{\star,0}$ solution to the collocation problem $-\Delta s_n = f$ at the Legendre nodes inside Ω, satisfies*

$$(1.3.4) \qquad \int_\Omega s_n^2 \, dx dy + \int_\Omega |\vec{\nabla} s_n|^2 dx dy \leq C \int_\Omega f^2 \, dx dy.$$

Proof - The two-dimensional counterpart of the inequality $(A.2.8)$ states that the quadrature and the integral are uniformly equivalent with respect to n, i.e. that for any $\phi \in \mathbf{P}_n^\star$ we have

$$(1.3.5) \qquad \int_\Omega \phi^2 \, dx dy \leq \sum_{i,j=0}^n \phi^2(\eta_j^{(n)}, \eta_i^{(n)}) \tilde{w}_i^{(n)} \tilde{w}_j^{(n)} \leq 9 \int_\Omega \phi^2 \, dx dy.$$

With the same arguments used for (1.3.2), i.e. quadrature formulas and integration by parts, the following relation holds true for any $\phi \in \mathbf{P}_n^{\star,0}$:

$$(1.3.6) \qquad \sum_{i,j=0}^n (\vec{\nabla} s_n \cdot \vec{\nabla} \phi)(\eta_j^{(n)}, \eta_i^{(n)}) \tilde{w}_i^{(n)} \tilde{w}_j^{(n)} = \sum_{i,j=0}^n (f\phi)(\eta_j^{(n)}, \eta_i^{(n)}) \tilde{w}_i^{(n)} \tilde{w}_j^{(n)}.$$

Now, we take $\phi = s_n$. The use of (1.3.5) and (1.3.6), together with the Schwarz inequality and (1.1.8), yields

$$(1.3.7) \qquad \int_\Omega |\vec{\nabla} s_n|^2 dx dy$$

$$\leq C_1 \left[\int_\Omega |\vec{\nabla} s_n|^2 dx dy \right]^{-1} \left[\sum_{i,j=0}^n |\vec{\nabla} s_n|^2 (\eta_j^{(n)}, \eta_i^{(n)}) \tilde{w}_i^{(n)} \tilde{w}_j^{(n)} \right]^2$$

$$\leq C_1 \left[\int_\Omega |\vec{\nabla} s_n|^2 dx dy \right]^{-1} \left\{ \left[\sum_{i,j=0}^n s_n^2 (\eta_j^{(n)}, \eta_i^{(n)}) \tilde{w}_i^{(n)} \tilde{w}_j^{(n)} \right] \right.$$

$$\left. \cdot \left[\sum_{i,j=0}^n f^2 (\eta_j^{(n)}, \eta_i^{(n)}) \tilde{w}_i^{(n)} \tilde{w}_j^{(n)} \right] \right\} \leq$$

$$\leq C_2 \sum_{i,j=0}^{n} (\tilde{I}_n f)^2 (\eta_j^{(n)}, \eta_i^{(n)}) \tilde{w}_i^{(n)} \tilde{w}_j^{(n)} \leq C_3 \int_\Omega (\tilde{I}_n f)^2 dx dy$$

$$\leq 2C_3 \left[\int_\Omega f^2 \, dx dy + \int_\Omega (\tilde{I}_n f - f)^2 \, dx dy \right] \leq C_4 \int_\Omega f^2 \, dx dy,$$

where C_1, C_2, C_3, C_4 are positive constants not depending on n. The last inequality is a consequence of (1.3.3). By (1.3.7) and (1.1.8) we also get a bound for $\int_\Omega s_n^2 dx dy$. The proof is concluded.

We next show a convergence result.

THEOREM 1.3.3 - *Let U be the solution of (1.1.1)-(1.1.2) with $g_k = 0$, $1 \leq k \leq 4$. Let $s_n \in \mathbf{P}_n^{\star,0}$, $n \geq 1$, be the solution to the collocation problem $-\Delta s_n = f$ at the Legendre nodes inside Ω. Then, provided U is sufficiently regular, one has*

$$(1.3.8) \qquad \lim_{n \to +\infty} \left(\int_\Omega (U - s_n)^2 dx dy + \int_\Omega |\vec{\nabla}(U - s_n)|^2 dx dy \right) = 0.$$

Proof - We define a polynomial $\chi_n \in \mathbf{P}_n^{\star,0}$ such that

$$(1.3.9) \qquad \int_\Omega (\vec{\nabla}\chi_n \cdot \vec{\nabla}\phi) dx dy = \int_\Omega (\vec{\nabla}U \cdot \vec{\nabla}\phi) dx dy, \qquad \forall \phi \in \mathbf{P}_n^{\star,0}.$$

Existence and uniqueness of χ_n are guaranteed by theorem 1.1.2 (the right-hand side of (1.3.9) can be viewed as a linear operator from $\mathbf{P}_n^{\star,0}$ to $\mathbf{R}$). The polynomial χ_n turns out to be the orthogonal projection of U into the space $\mathbf{P}_n^{\star,0}$ in a suitable inner product.
At this point, we assume that

$$(1.3.10) \qquad \lim_{n \to +\infty} \int_\Omega |\vec{\nabla}(U - \chi_n)|^2 dx dy = 0,$$

$$(1.3.11) \qquad \lim_{n \to +\infty} \sup_{\phi \in \mathbf{P}_n^{\star,0}} \left\{ \left[\int_\Omega |\vec{\nabla}\phi|^2 dx dy \right]^{-\frac{1}{2}} \right.$$

$$\left. \cdot \left[\int_\Omega f\phi \, dx dy - \sum_{i,j=0}^{n} (f\phi)(\eta_j^{(n)}, \eta_i^{(n)}) \tilde{w}_i^{(n)} \tilde{w}_j^{(n)} \right] \right\} = 0,$$

$$(1.3.12) \qquad \lim_{n \to +\infty} \sup_{\phi \in \mathbf{P}_n^{*,0}} \left\{ \left[\int_{\Omega} |\vec{\nabla}\phi|^2 dx dy \right]^{-\frac{1}{2}} \right.$$

$$\left. \left[\int_{\Omega} (\vec{\nabla}\chi_n \cdot \vec{\nabla}\phi) dx dy - \sum_{i,j=0}^{n} (\vec{\nabla}\chi_n \cdot \vec{\nabla}\phi)(\eta_j^{(n)}, \eta_i^{(n)}) \tilde{w}_i^{(n)} \tilde{w}_j^{(n)} \right] \right\} = 0,$$

which hold true, provided f and U are sufficiently regular functions. The proof of (1.3.10) is found in CANUTO and QUARTERONI (1982). The limits (1.3.11) and (1.3.12) are a consequence of the convergence of the Gaussian quadrature formulas to the exact integrals. Estimates are given in BRESSAN and QUARTERONI (1986) in the Chebyshev case where the integrals are weighted by the function $\omega(x) := (1-x^2)^{-1/2}(1-y^2)^{-1/2}$ (the Legendre case is trivially obtained by taking $\omega = 1$). In all the limits the rate of convergence depends on the regularity of functions f and U.

The conclusion of our theorem is now straightforward. In fact, by the triangle inequality we have

$$(1.3.13) \qquad \left[\int_{\Omega} |\vec{\nabla}(U - s_n)|^2 dx dy \right]^{\frac{1}{2}}$$

$$\leq \left[\int_{\Omega} |\vec{\nabla}(U - \chi_n)|^2 dx dy \right]^{\frac{1}{2}} + \left[\int_{\Omega} |\vec{\nabla}\phi|^2 dx dy \right]^{\frac{1}{2}},$$

where we defined $\phi := \chi_n - s_n \in \mathbf{P}_n^{*,0}$. Due to (1.3.10), the first term on the right-hand side of (1.3.13) tends to zero. Thanks to (1.3.5), (1.3.6), (1.1.6) and (1.3.9) the second term is estimated as follows

$$(1.3.14) \qquad \left[\int_{\Omega} |\vec{\nabla}\phi|^2 dx dy \right]^{\frac{1}{2}}$$

$$\leq C \left[\int_{\Omega} |\vec{\nabla}\phi|^2 dx dy \right]^{-\frac{1}{2}} \left[\sum_{i,j=0}^{n} |\vec{\nabla}\phi|^2 (\eta_j^{(n)}, \eta_i^{(n)}) \tilde{w}_i^{(n)} \tilde{w}_j^{(n)} \right]$$

$$= C \left[\int_{\Omega} |\vec{\nabla}\phi|^2 dx dy \right]^{-\frac{1}{2}} \left[\sum_{i,j=0}^{n} (\vec{\nabla}\chi_n \cdot \vec{\nabla}\phi)(\eta_j^{(n)}, \eta_i^{(n)}) \tilde{w}_i^{(n)} \tilde{w}_j^{(n)} \right.$$

$$\left. - \sum_{i,j=0}^{n} (f\phi)(\eta_j^{(n)}, \eta_i^{(n)}) \tilde{w}_i^{(n)} \tilde{w}_j^{(n)} \right] =$$

$$= C \left[\int_\Omega |\vec{\nabla}\phi|^2 \, dx \, dy \right]^{-\frac{1}{2}} \left[\int_\Omega f\phi \, dx \, dy - \sum_{i,j=0}^{n} (f\phi)(\eta_j^{(n)}, \eta_i^{(n)}) \tilde{w}_i^{(n)} \tilde{w}_j^{(n)} \right]$$

$$+ C \left[\int_\Omega |\vec{\nabla}\phi|^2 \, dx \, dy \right]^{-\frac{1}{2}} \left[\sum_{i,j=0}^{n} (\vec{\nabla}\chi_n \cdot \vec{\nabla}\phi)(\eta_j^{(n)}, \eta_i^{(n)}) \tilde{w}_i^{(n)} \tilde{w}_j^{(n)} \right.$$

$$\left. - \int_\Omega (\vec{\nabla}\chi_n \cdot \vec{\nabla}\phi) \, dx \, dy \right],$$

where $C > 0$ does not depend on n. Recalling (1.3.11) and (1.3.12), the above quantities tend to zero. A bound for the error $\int_\Omega (U - s_n)^2 \, dx \, dy$ is then obtained by virtue of (1.1.8). Therefore, we have proved (1.3.8).

The rate of convergence in (1.3.8) is determined by the regularity of U. Actually the error behaves as: $\dfrac{1}{n^{\sigma-1}} \displaystyle\sum_{\alpha+\beta=\sigma} \int_\Omega \left(\dfrac{\partial^\sigma}{\partial x^\alpha \partial y^\beta} U \right)^2 \, dx \, dy$, where $\sigma > 1$. Hence, for a very regular solution U, the error decays faster than any fixed power of n^{-1}, which is the peculiarity of spectral methods.

Similar results can be also given for non-homogeneous boundary conditions (see MADAY (1989)), although the proof is more technical. In particular, theorem 1.3.2 is generalized by adding to the right-hand side of (1.3.4) a suitable norm of the functions g_k, $1 \le k \le 4$, which are required to satisfy the compatibility conditions: $g_1(1) = g_2(-1)$, $g_2(1) = g_3(1)$, $g_3(-1) = g_4(1)$, $g_4(-1) = g_1(-1)$.

1.4 Numerical implementation of the collocation method

We turn our attention to the numerical solution of the linear system (1.2.6). We recall that the matrix A_n has dimension $(n+1)^2$, and, although there are many entries equal to zero (see for instance Fig. 1.2.3), it does not have a band structure. Hence, the use of a direct solver, such as the one based on the *LU* factorization (that is, the decomposition of A_n by lower and upper triangular matrices), would have a cost proportional to n^6, which is too high in view of practical applications.

We note that the evaluation of a partial derivative in $\mathbf{P}_n^*$ corresponds to multiplication $n + 1$ times by the $(n + 1) \times (n + 1)$ matrix $\tilde{D}_n$ representing the one-dimensional derivative operator (see section A.3). Therefore, instead of n^4, the cost of the multiplication of $\mathcal{A}_n$ by a vector is only proportional to n^3 (this is also evident after examining (1.2.4)). It is also worthwhile to remember that, according to theorem 1.3.1, the eigenvalues of $\mathcal{A}_n$ are real and positive. Consequently, the use of an iterative solver seems to be a practicable alternative to the direct solver. In this way we save memory, since we do not need to assemble the whole matrix $\mathcal{A}_n$, and we can construct and store once and for all the smaller matrices $\tilde{D}_n$ and $\tilde{D}_n^2$ (first and second derivative operators, respectively).

There is, however, a serious drawback: $\mathcal{A}_n$ is very *ill-conditioned*. By ill-conditioning we mean that the ratio between the biggest and the smallest eigenvalues of $\mathcal{A}_n$ is very large. Actually, the minimum eigenvalue converges with spectral accuracy to the minimum eigenvalue of the operator $-\Delta$ in Ω, which is $\frac{1}{2}\pi^2$, while the maximum eigenvalue is proportional to n^4. Such a discrepancy makes the convergence of the standard iterative methods very slow, so that the total computer time needed is still too much. The only way to realize a fast computational code is to adopt a *preconditioner*, i.e. an invertible matrix $\mathcal{B}_n$ of the same dimension of $\mathcal{A}_n$, such that $\mathcal{B}_n^{-1}$ can be cheaply evaluated, and the eigenvalues of $\mathcal{B}_n^{-1}\mathcal{A}_n$ are close together. We are now going to show how to construct a suitable preconditioner.

For any fixed $n \geq 1$, we consider the finite-differences discretization of the equation (1.1.1)-(1.1.2) by a centered five-points scheme based on the Legendre grid $\Theta_k^{(n)}$, $0 \leq k \leq n_T$.
This amounts to determining $n_T + 1$ unknowns $\{z_{i,j}\}_{\substack{0 \leq i \leq n \\ 0 \leq j \leq n}}$ such that

$$
(1.4.1) \quad -\frac{2}{\eta_{j+1}^{(n)} - \eta_{j-1}^{(n)}} \left(\frac{z_{i,j+1} - z_{i,j}}{\eta_{j+1}^{(n)} - \eta_j^{(n)}} - \frac{z_{i,j} - z_{i,j-1}}{\eta_j^{(n)} - \eta_{j-1}^{(n)}} \right)
$$

$$
-\frac{2}{\eta_{i+1}^{(n)} - \eta_{i-1}^{(n)}} \left(\frac{z_{i+1,j} - z_{i,j}}{\eta_{i+1}^{(n)} - \eta_i^{(n)}} - \frac{z_{i,j} - z_{i-1,j}}{\eta_i^{(n)} - \eta_{i-1}^{(n)}} \right) = f(\eta_j^{(n)}, \eta_i^{(n)}),
$$

for $1 \leq i \leq n - 1$, $1 \leq j \leq n - 1$. In addition, we must impose the boundary constraints

$$
(1.4.2) \quad
\begin{cases}
z_{j,0} = g_1(\eta_j^{(n)}) & 0 \leq j \leq n - 1, \\[2mm]
z_{n,i} = g_2(\eta_i^{(n)}) & 0 \leq i \leq n - 1, \\[2mm]
z_{j,n} = g_3(\eta_j^{(n)}) & 1 \leq j \leq n, \\[2mm]
z_{0,i} = g_4(\eta_i^{(n)}) & 1 \leq i \leq n.
\end{cases}
$$

$$
\begin{bmatrix}
1 & 0 & 0 & 0 & 0 & 0 & 0 & 0 & 0 & 0 & 0 & 0 & 0 & 0 & 0 & 0 & 0 & 0 & 0 & 0 & 0 & 0 & 0 & 0 & 0 & 0 & 0 \\
0 & 1 & 0 & 0 & 0 & 0 & 0 & 0 & 0 & 0 & 0 & 0 & 0 & 0 & 0 & 0 & 0 & 0 & 0 & 0 & 0 & 0 & 0 & 0 & 0 & 0 & 0 \\
0 & 0 & 1 & 0 & 0 & 0 & 0 & 0 & 0 & 0 & 0 & 0 & 0 & 0 & 0 & 0 & 0 & 0 & 0 & 0 & 0 & 0 & 0 & 0 & 0 & 0 & 0 \\
0 & 0 & 0 & 1 & 0 & 0 & 0 & 0 & 0 & 0 & 0 & 0 & 0 & 0 & 0 & 0 & 0 & 0 & 0 & 0 & 0 & 0 & 0 & 0 & 0 & 0 & 0 \\
0 & 0 & 0 & 0 & 1 & 0 & 0 & 0 & 0 & 0 & 0 & 0 & 0 & 0 & 0 & 0 & 0 & 0 & 0 & 0 & 0 & 0 & 0 & 0 & 0 & 0 & 0 \\
0 & 0 & 0 & 0 & 0 & 1 & 0 & 0 & 0 & 0 & 0 & 0 & 0 & 0 & 0 & 0 & 0 & 0 & 0 & 0 & 0 & 0 & 0 & 0 & 0 & 0 & 0 \\
0 & \bullet & 0 & 0 & 0 & \bullet & \bullet & \bullet & 0 & 0 & 0 & \bullet & 0 & 0 & 0 & 0 & 0 & 0 & 0 & 0 & 0 & 0 & 0 & 0 & 0 & 0 & 0 \\
0 & 0 & \bullet & 0 & 0 & 0 & \bullet & \bullet & \bullet & 0 & 0 & 0 & \bullet & 0 & 0 & 0 & 0 & 0 & 0 & 0 & 0 & 0 & 0 & 0 & 0 & 0 & 0 \\
0 & 0 & 0 & \bullet & 0 & 0 & 0 & \bullet & \bullet & \bullet & 0 & 0 & 0 & \bullet & 0 & 0 & 0 & 0 & 0 & 0 & 0 & 0 & 0 & 0 & 0 & 0 & 0 \\
0 & 0 & 0 & 0 & 0 & 0 & 0 & 0 & 0 & 1 & 0 & 0 & 0 & 0 & 0 & 0 & 0 & 0 & 0 & 0 & 0 & 0 & 0 & 0 & 0 & 0 & 0 \\
0 & 0 & 0 & 0 & 0 & 0 & 0 & 0 & 0 & 0 & 1 & 0 & 0 & 0 & 0 & 0 & 0 & 0 & 0 & 0 & 0 & 0 & 0 & 0 & 0 & 0 & 0 \\
0 & 0 & 0 & 0 & 0 & 0 & \bullet & 0 & 0 & 0 & \bullet & \bullet & \bullet & \bullet & 0 & 0 & 0 & \bullet & 0 & 0 & 0 & 0 & 0 & 0 & 0 & 0 & 0 \\
0 & 0 & 0 & 0 & 0 & 0 & 0 & \bullet & 0 & 0 & 0 & \bullet & \bullet & \bullet & \bullet & 0 & 0 & 0 & \bullet & 0 & 0 & 0 & 0 & 0 & 0 & 0 & 0 \\
0 & 0 & 0 & 0 & 0 & 0 & 0 & 0 & \bullet & 0 & 0 & 0 & \bullet & \bullet & \bullet & \bullet & 0 & 0 & 0 & \bullet & 0 & 0 & 0 & 0 & 0 & 0 & 0 \\
0 & 0 & 0 & 0 & 0 & 0 & 0 & 0 & 0 & 0 & 0 & 0 & 0 & 0 & 0 & 1 & 0 & 0 & 0 & 0 & 0 & 0 & 0 & 0 & 0 & 0 & 0 \\
0 & 0 & 0 & 0 & 0 & 0 & 0 & 0 & 0 & 0 & 0 & 0 & 0 & 0 & 0 & 0 & 1 & 0 & 0 & 0 & 0 & 0 & 0 & 0 & 0 & 0 & 0 \\
0 & 0 & 0 & 0 & 0 & 0 & 0 & 0 & 0 & 0 & 0 & \bullet & 0 & 0 & 0 & \bullet & \bullet & \bullet & \bullet & 0 & 0 & 0 & \bullet & 0 & 0 & 0 & 0 \\
0 & 0 & 0 & 0 & 0 & 0 & 0 & 0 & 0 & 0 & 0 & 0 & \bullet & 0 & 0 & 0 & \bullet & \bullet & \bullet & \bullet & 0 & 0 & 0 & \bullet & 0 & 0 & 0 \\
0 & 0 & 0 & 0 & 0 & 0 & 0 & 0 & 0 & 0 & 0 & 0 & 0 & \bullet & 0 & 0 & 0 & \bullet & \bullet & \bullet & \bullet & 0 & 0 & 0 & \bullet & 0 & 0 \\
0 & 0 & 0 & 0 & 0 & 0 & 0 & 0 & 0 & 0 & 0 & 0 & 0 & 0 & 0 & 0 & 0 & 0 & 0 & 0 & 1 & 0 & 0 & 0 & 0 & 0 & 0 \\
0 & 0 & 0 & 0 & 0 & 0 & 0 & 0 & 0 & 0 & 0 & 0 & 0 & 0 & 0 & 0 & 0 & 0 & 0 & 0 & 0 & 1 & 0 & 0 & 0 & 0 & 0 \\
0 & 0 & 0 & 0 & 0 & 0 & 0 & 0 & 0 & 0 & 0 & 0 & 0 & 0 & 0 & 0 & 0 & 0 & 0 & 0 & 0 & 0 & 1 & 0 & 0 & 0 & 0 \\
0 & 0 & 0 & 0 & 0 & 0 & 0 & 0 & 0 & 0 & 0 & 0 & 0 & 0 & 0 & 0 & 0 & 0 & 0 & 0 & 0 & 0 & 0 & 1 & 0 & 0 & 0 \\
0 & 0 & 0 & 0 & 0 & 0 & 0 & 0 & 0 & 0 & 0 & 0 & 0 & 0 & 0 & 0 & 0 & 0 & 0 & 0 & 0 & 0 & 0 & 0 & 1 & 0 & 0 \\
0 & 0 & 0 & 0 & 0 & 0 & 0 & 0 & 0 & 0 & 0 & 0 & 0 & 0 & 0 & 0 & 0 & 0 & 0 & 0 & 0 & 0 & 0 & 0 & 0 & 1 & 0 \\
0 & 0 & 0 & 0 & 0 & 0 & 0 & 0 & 0 & 0 & 0 & 0 & 0 & 0 & 0 & 0 & 0 & 0 & 0 & 0 & 0 & 0 & 0 & 0 & 0 & 0 & 1
\end{bmatrix}
$$

FIG. 1.4.1 - *Structure of the preconditioning matrix* $\mathcal{B}_4$.

In other words, the set of equations in (1.4.1) is equivalent to

$$
(1.4.3) \qquad -(\Delta Q_{i,j})(\eta_j^{(n)}, \eta_i^{(n)}) = f(\eta_j^{(n)}, \eta_i^{(n)}),
$$

where $Q_{i,j}$ is the unique polynomial of degree 2 in each variable (thereby depending on 9 degrees of freedom) which attains the values $z_{i+\alpha,j+\beta}$ at the nine points $(\eta_{j+\beta}^{(n)}, \eta_{i+\alpha}^{(n)})$, respectively, where α and β belong to the set $\{-1, 0, 1\}$ (i.e. the points A_k, $1 \le k \le 9$, of Fig. 2.2.1).

We organize the vector of the unknowns $\vec{Z}_n$ using the same ordering adopted for the nodes in (1.2.3). Afterwards, we denote by $\mathcal{B}_n$ the matrix corresponding to the linear system (1.4.1)-(1.4.2). This leads to the equation $\mathcal{B}_n \vec{Z}_n = \vec{B}_n$, where the right-hand side is the same as the one of equation (1.2.6). It is not difficult to recognize that $z_{i,j}$ is an approximation of $U(\eta_j^{(n)}, \eta_i^{(n)})$. Neverthe-less, even if we are using the Legendre nodes, the finite-difference discretization does not have the same convergence rate as the spectral collocation method, which is much more accurate. This aspect however is not interesting to us, since we are more concerned with some other properties of the matrix $\mathcal{B}_n$. Actually, $\mathcal{B}_n$ is non-symmetric, with real and positive eigenvalues and it is banded with bandwidth $2n + 3$. We can see the structure of $\mathcal{B}_4$ in Fig. 1.4.1.

The cost of computing the LU factorization of $\mathcal{B}_n$ is proportional to n^4, when taking into account its structure. In addition, the factorization process only requires a memory storage of the size of the band. After factorization, the multiplication of $\mathcal{B}_n^{-1} = U^{-1}L^{-1}$ by a vector can be carried out with a cost proportional to n^3.

We claim that $\mathcal{B}_n$ is a good preconditioner for $\mathcal{A}_n$, as first pointed out in ORSZAG (1980). There is, in fact, numerical evidence that the eigenvalues of $\mathcal{B}_n^{-1}\mathcal{A}_n$ are real and lie in the interval $[1, \frac{1}{4}\pi^2[$. In the first column of Table 1.4.1, we report the maximum eigenvalue $\lambda_{\max}^{(n)}$ of $\mathcal{B}_n^{-1}\mathcal{A}_n$ for various n.

n	$\lambda_{\max}^{(n)}$	$\mu^{(n)}$	$\rho^{(n)}$
4	1.554	1.668	0.109
8	1.945	2.045	0.164
12	2.101	2.180	0.183
16	2.185	2.251	0.193
20	2.237	2.294	0.198

TABLE 1.4.1 - *Exact and estimated maximum eigenvalues of* $\mathcal{B}_n^{-1}\mathcal{A}_n$.

Consequently, the system $\mathcal{B}_n^{-1}\mathcal{A}_n\vec{X}_n = \mathcal{B}_n^{-1}\vec{B}_n$ turns out to have the same solution as (1.2.6), but the distribution of the eigenvalues of the preconditioned matrix $\mathcal{B}_n^{-1}\mathcal{A}_n$ is now uniformly bounded with respect to n.

Due to the imposition of the boundary conditions, we realize that the eigen-value $\lambda_{\min}^{(n)} = 1$ has multiplicity of at least $4n$. There is another polynomial eigenfunction corresponding to the eigenvalue 1, i.e. $b(x,y) := (1-x^2)(1-y^2)$. This is true because $\mathcal{A}_n$ and $\mathcal{B}_n$ coincide when restricted to the subspace $\mathbf{P}_2^{\star,0} \subset \mathbf{P}_n^{\star,0}$. The function b is a *low eigenmode* (see BOYD (1989), section 12.4). *High eigenmodes* are related to large eigenvalues and they usually present os-cillations at high frequency, with a wavelength of an order of the distance between the grid points. We are unable to give an explicit expression to the highest eigenmodes of $B_n^{-1}A_n$. However, we can get fairly close to $\lambda_{\max}^{(n)}$ by applying the following procedure, which will be also useful in section 2.3.

We take $\phi_n \in \mathbf{P}_n^\star$ such that $\phi_n(\eta_j^{(n)}, \eta_i^{(n)}) := (-1)^{i+j}$, which corresponds to a vector of the form $\vec{\Phi}_n \equiv (1, -1, 1, -1, \cdots, -1, 1)$. Then, we compute $\vec{\Psi}_n := B_n^{-1}A_n\vec{\Phi}_n$, obtaining a new polynomial $\psi_n \in \mathbf{P}_n^\star$ ($\vec{\Psi}_n$ is the vector of the values of ψ_n at the Legendre grid). We finally consider the ratio

$$(1.4.4) \qquad \mu^{(n)} := \frac{\|\vec{\Psi}_n\|}{\|\vec{\Phi}_n\|},$$

where the norm $\|\cdot\|$ in $\mathbf{R}^{(n+1)^2}$ is defined by

$$(1.4.5) \qquad \|\vec{\Phi}_n\| := \sqrt{\int_\Omega \phi_n^2\, dx dy} = \left(\sum_{i,j=0}^n \phi_n^2(\eta_j^{(n)}, \eta_i^{(n)})\, \tilde{w}_i^{(n)} \tilde{w}_j^{(n)} \right.$$

$$\left. - \frac{n(n+1)}{2(2n+1)} \sum_{i,j=0}^n \phi_n(\eta_j^{(n)}, \eta_i^{(n)}) \left[P_n(\eta_i^{(n)}) + P_n(\eta_j^{(n)}) \right] \tilde{w}_i^{(n)} \tilde{w}_j^{(n)} \right)^{1/2}$$

corresponding to the two-dimensional version of formula $(A.2.7)$.

The quantity $\mu^{(n)}$ is basically the result after one step of the *power method* to evaluate the maximum eigenvalue of a given matrix. As shown in Table 1.4.1, $\mu^{(n)}$ is an estimate of $\lambda_{\max}^{(n)}$ taken from above. Therefore, we can conjecture that the eigenvalues of $B_n^{-1}A_n$ belong to the interval $[1, \mu^{(n)}[$.

We are left with the problem of devising a suitable iterative technique for the solution of (1.2.6). We suggest the Du Fort-Frankel method (see DU FORT and FRANKEL (1953), RICHTMYER and MORTON (1967), JAIN (1984)). From a couple of initial vectors $\vec{X}_n^0$ and $\vec{X}_n^1$, one constructs by recursion the sequence

$$(1.4.6) \qquad \vec{X}_n^{k+1} = \frac{2\sigma_1}{1+2\sigma_1\sigma_2} \left[-B_n^{-1}(A_n\vec{X}_n^k - \vec{B}_n) + 2\sigma_2\vec{X}_n^k \right]$$

$$+ \frac{1-2\sigma_1\sigma_2}{1+2\sigma_1\sigma_2} \vec{X}_n^{k-1},$$

where $k \geq 1$ and σ_1, σ_2 are positive parameters. The method converges and $\lim_{k\to+\infty} \vec{X}_n^k = \vec{X}_n$, provided σ_1 and σ_2 satisfy some hypotheses depending on the distribution of the eigenvalues of $B_n^{-1}A_n$. The explicit Du Fort-Frankel two-step scheme has a low cost of implementation and a good rate of convergence. Moreover, it does not require the matrices to be symmetric, as many other methods with equivalent performances do.

The best convergence rate is achieved with the optimal parameters

$$(1.4.7) \qquad \sigma_1 = [\lambda_{\max}^{(n)} \lambda_{\min}^{(n)}]^{-\frac{1}{2}}, \qquad \sigma_2 = \tfrac{1}{4}(\lambda_{\max}^{(n)} + \lambda_{\min}^{(n)}),$$

$\lambda_{\max}^{(n)}$ and $\lambda_{\min}^{(n)}$ being the maximum and the minimum eigenvalues of $\mathcal{B}_n^{-1}\mathcal{A}_n$ respectively. With σ_1 and σ_2 as in (1.4.7), the *spectral radius* of the *amplification matrix* associated to (1.4.6) is

$$(1.4.8) \qquad \rho^{(n)} := \frac{\sqrt{\lambda_{\max}^{(n)}/\lambda_{\min}^{(n)}} - 1}{\sqrt{\lambda_{\max}^{(n)}/\lambda_{\min}^{(n)}} + 1}.$$

In Table 1.4.1 we find the values of $\rho^{(n)}$ for some n. We recall that $\rho^{(n)}$ must be less than 1 for convergence. In our case $\rho^{(n)} \ll 1$, allowing a very fast damping of the error $\vec{X}_n^k - \vec{X}_n$ in a number of iterations not depending on n. Since $\lambda_{\min}^{(n)} = 1$, and $\lambda_{\max}^{(n)}$ is not known in general, we choose the parameters as follows

$$(1.4.9) \qquad \sigma_1 = [\mu^{(n)}]^{-\frac{1}{2}}, \qquad \sigma_2 = \tfrac{1}{4}(\mu^{(n)} + 1),$$

where $\mu^{(n)}$ is given by (1.4.4). These values are not too far from the optimal ones, and they also insure a good convergence rate. As initial guesses, it turns out to be convenient to take $\vec{X}_n^0 \equiv \vec{X}_n^1 \equiv \vec{Z}_n$, where $\vec{Z}_n$ is the solution to $\mathcal{B}_n \vec{Z}_n = \vec{B}_n$.

The whole approximation procedure can be generalized to the case in which different polynomial degrees are taken in the x and y directions. The convergence theory of section 1.3 can be easily adapted to the new situation. We remind the reader that now the bandwidth of the preconditioner varies depending on whether the nodes in Ω are ordered by rows or columns. Of course, the less expensive choice will be the one minimizing the size of the band.

We now have all the elements to build up a fast and reliable numerical code to find the approximated solution of the Poisson problem by the Legendre collocation method. We will summarize the main steps of the algorithm in the next section.

1.5   The numerical algorithm for the Poisson equation

The procedure for the numerical solution of the system (1.2.6) is described by the following flow-chart. We first need some initialization steps:

1) Compute the nodes $\eta_i^{(n)}$, $0 \le i \le n$, and the weights $\tilde{w}_i^{(n)}$, $0 \le i \le n$ (see section A.2);

2) Compute the entries of the differentiation matrices $\tilde{D}_n$ and $\tilde{D}_n^2 = \tilde{D}_n \tilde{D}_n$ (see section A.3);

3) Construct the banded matrix $\mathcal{B}_n$;

4) Using the structure of $\mathcal{B}_n$, compute its LU factorization;

5) Construct the right-hand side vector $\vec{B}_n$ (see section 1.2);

6) Solve the system $\mathcal{B}_n \vec{Z}_n = \vec{B}_n$ with the help of the LU decomposition of $\mathcal{B}_n$;

7) Set $\vec{X}_n^0 \equiv \vec{X}_n^1 \equiv \vec{Z}_n$ to be the starting vectors;

8) Use the vectors $\vec{\Phi}_n$ and $\vec{\Psi}_n$ as suggested in section 1.4 to determine $\mu^{(n)}$ in (1.4.4) and the values of σ_1 and σ_2 in (1.4.9).

Then, for $k \geq 1$, we advance with the loop:

9) Compute $\mathcal{A}_n \vec{X}_n^k$ by repeated applications of the matrix $\tilde{D}_n^2$ (see (1.2.4));

10) Compute $\mathcal{B}_n^{-1}(\mathcal{A}_n \vec{X}_n^k - \vec{B}_n)$ using the LU decomposition of $\mathcal{B}_n$;

11) Update $\vec{X}_n^{k+1}$ according to (1.4.6), with σ_1 and σ_2 as chosen in step 8;

12) Check the norm of the residual: $\|\mathcal{A}_n \vec{X}_n^{k+1} - \vec{B}_n\|$, where $\|\cdot\|$ is either defined in (1.4.5) or is another equivalent norm. If this is less than a prescribed error then stop, otherwise go back to step 9.

The resulting output vector is the approximation of the solution U of (1.1.1)-(1.1.2) at the Legendre grid $(\eta_j^{(n)}, \eta_i^{(n)})$, $0 \leq i \leq n$, $0 \leq j \leq n$. The convergence to zero of the errors $\|\mathcal{A}_n \vec{X}_n^k - \vec{B}_n\|$ and $\|\vec{X}_n^k - \vec{X}_n\|$ is monotone, so that the stopping test at step 12 is a reliable criterion to check that the method has reached its steady state.

———— ◇ ————

As an example, we study the results corresponding to the approximation of the following partial differential equation

$$(1.5.1) \qquad \begin{cases} -\Delta U = 1 & \text{in } \Omega, \\ U = 0 & \text{on } \partial\Omega. \end{cases}$$

We show in Fig. 1.5.1 the contour lines of U. For any $n \geq 1$, we denote by $q_n \in \mathbf{P}_n^{\star,0}$ the polynomial solution to the corresponding collocation problem. Then, we generate the sequence of vectors $\vec{X}_n^k$ by the Du Fort-Frankel method, and we denote by $q_n^{(k)} \in \mathbf{P}_n^{\star,0}$, $k \geq 0$, the sequence of polynomials such that $q_n^{(k)}$ attains the values $\vec{X}_n^k$ at the Legendre grid.

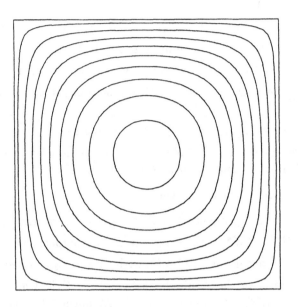

FIG. 1.5.1 - *The solution to problem* (1.5.1).

k	$n = 4$	$n = 8$	$n = 12$	$n = 16$	$n = 20$
2	$.137E - 1$	$.149E - 1$	$.110E - 1$	$.784E - 2$	$.578E - 2$
3	$.111E - 2$	$.206E - 2$	$.222E - 2$	$.178E - 2$	$.140E - 2$
4	$.315E - 3$	$.204E - 3$	$.390E - 3$	$.370E - 3$	$.313E - 3$
5	$.813E - 5$	$.296E - 4$	$.606E - 4$	$.710E - 4$	$.656E - 4$
6	$.443E - 5$	$.103E - 4$	$.818E - 5$	$.128E - 4$	$.132E - 4$
7	$.499E - 6$	$.244E - 5$	$.104E - 5$	$.214E - 5$	$.253E - 5$
8	$.305E - 7$	$.358E - 6$	$.214E - 6$	$.334E - 6$	$.469E - 6$
9	$.106E - 7$	$.350E - 7$	$.624E - 7$	$.475E - 7$	$.822E - 7$
10	$.385E - 9$	$.781E - 8$	$.158E - 7$	$.788E - 8$	$.139E - 7$

TABLE 1.5.1 - *Norms of the residuals in the approximation of problem* (1.5.1).

In Table 1.5.1, for various n and k, we examine the norm, determined according to (1.4.5), of the residual $r_n^{(k)} \in \mathbf{P}_n^{\star,0}$ which is characterized by the set of relations

$$(1.5.2) \qquad r_n^{(k)}(\eta_j^{(n)}, \eta_i^{(n)}) := (-\Delta q_n^{(k)} - 1)(\eta_j^{(n)}, \eta_i^{(n)}),$$

for $1 \le i \le n-1$, $1 \le j \le n-1$. As the reader will notice, the convergence is very fast, and no more than 15 iterations are usually needed to reach the machine accuracy in double precision. We recall that the cost of the algorithm is proportional to n^4 for what concerns the factorization of the preconditioner, while each iteration has a cost proportional to n^3. We observe that, due to the ill-conditioning of $\mathcal{A}_n$, the error $\|\vec{X}_n^k - \vec{X}_n\|$ is even smaller than the error of the residual. The approximated solutions q_n, $n \ge 2$, look accurate and a spectral convergence rate is observed, although there is a loss of regularity of the function U at the corner points. Similar experiments have been reported in HAIDVOGEL and ZANG (1979), and CANUTO and QUARTERONI (1985).

1.6 Application to the eikonal equation

In this section, we are concerned with the solution $U_\epsilon : \bar{\Omega} \to \mathbf{R}$ to the nonlinear equation

$$(1.6.1) \qquad \begin{cases} -\epsilon \, \Delta U_\epsilon + \sqrt{\left(\frac{\partial}{\partial x} U_\epsilon\right)^2 + \left(\frac{\partial}{\partial y} U_\epsilon\right)^2} = 1 & \text{in } \Omega, \\ U_\epsilon = 0 & \text{on } \partial\Omega, \end{cases}$$

where $\Omega =]-1, 1[\times]-1, 1[$ and $\epsilon > 0$ is a given parameter.

Theoretical analysis says that (1.6.1) admits a unique solution. For $\epsilon \to 0$ one may also prove that U_ϵ converges uniformly to the function $U : \bar{\Omega} \to \mathbf{R}$ defined by

$$(1.6.2) \qquad U(x, y) := 1 - \max\{|x|, |y|\}, \qquad (x, y) \in \bar{\Omega},$$

which satisfies

$$(1.6.3) \qquad \begin{cases} |\vec{\nabla} U| = 1 & \text{almost everywhere in } \Omega, \\ U = 0 & \text{on } \partial\Omega. \end{cases}$$

The contour lines of U are shown in Fig. 1.6.1. The equation (1.6.3) is known as *eikonal equation* (see COURANT and HILBERT (1953), WHITHAM (1974), ZAUDERER (1983) and BLEISTEIN (1984)), and it is related for example to the evolution of wave fronts in *geometrical optics* (in this case the light rays are normal to the fronts). We remark that the function U in (1.6.2) is not a unique solution to (1.6.3), but it is the one with a physical meaning.

The collocation method applied to equation (1.6.1) consists in finding $q_{\epsilon,n} \in \mathbf{P}_n^{\star,0}$ such that

$$(1.6.4) \qquad \left[-\epsilon \, \Delta q_{\epsilon,n} + |\vec{\nabla} q_{\epsilon,n}| \right] (\eta_j^{(n)}, \eta_i^{(n)}) = 1,$$

for $1 \leq i \leq n-1$, $1 \leq j \leq n-1$. The nonlinear set of equations in (1.6.4) is successively approximated by constructing a sequence of polynomials $q_{\epsilon,n}^{(k)} \in \mathbf{P}_n^{\star,0}$, $k \in \mathbf{N}$, determined by the recursion relation

$$(1.6.5) \qquad \frac{q_{\epsilon,n}^{(k+1)} - q_{\epsilon,n}^{(k)}}{\delta} - \epsilon \, \Delta q_{\epsilon,n}^{(k+1)} + |\vec{\nabla} q_{\epsilon,n}^{(k)}| = 1,$$

at the nodes $(\eta_j^{(n)}, \eta_i^{(n)})$, $1 \leq i \leq n-1$, $1 \leq j \leq n-1$, where $\delta > 0$ and $q_{\epsilon,n}^{(0)}$ are assigned. For sufficiently small δ the algorithm (1.6.5) converges and $\lim_{k \to +\infty} q_{\epsilon,n}^{(k)} = q_{\epsilon,n}$. The scheme is implicit for the Laplace operator in order to avoid severe restrictions on δ (see also section 5.1 where algorithms for other nonlinear equations are considered).

Denoting by $\vec{X}_{\epsilon,n}^k$ the vector of the values of $q_{\epsilon,n}^{(k)}$ at the Legendre grid, (1.6.5) is equivalent to solve the sequence of linear systems

$$(1.6.6) \qquad (\mathcal{I}_n + \epsilon \delta \mathcal{A}_n) \vec{X}_{\epsilon,n}^{k+1} = \vec{X}_{\epsilon,n}^k + \vec{\Phi}_{\epsilon,n}^k,$$

where $\mathcal{I}_n$ is the $(n+1)^2 \times (n+1)^2$ identity matrix and $\mathcal{A}_n$ is the discretization of the Laplace operator with boundary conditions introduced in section 1.2. On the right-hand side we find the vector $\vec{\Phi}_{\epsilon,n}^k$ of the values attained at the nodes by the polynomial $\phi_{\epsilon,n}^{(k)} \in \mathbf{P}_n^{\star,0}$ defined by

$$(1.6.7) \qquad \phi_{\epsilon,n}^{(k)}(\eta_j^{(n)}, \eta_i^{(n)}) = \delta \left[1 - |\vec{\nabla} q_{\epsilon,n}^{(k)}| \right] (\eta_j^{(n)}, \eta_i^{(n)}),$$

with $1 \leq i \leq n-1$, $1 \leq j \leq n-1$. For the numerical solution of (1.6.6), we adopt the preconditioned iterative algorithm introduced in sections 1.4 and 1.5, using as preconditioner the banded matrix $\mathcal{I}_n + \epsilon \delta \mathcal{B}_n$. The preconditioning matrix does not depend on k. Hence, it is factorized before proceeding with the iterative method.

We now give the results of some numerical tests. We start by taking $\epsilon = .05$ and $n = 16$. The plot of the corresponding polynomial $q_{\epsilon,n}$ is shown in Fig. 1.6.2. When ϵ is smaller (say for instance $\epsilon = .025$), the same polynomial degree is not sufficient to recover an accurate approximation of the exact solution U_ϵ (Fig. 1.6.3). However, the situation improves by choosing $n = 20$ (Fig. 1.6.4).

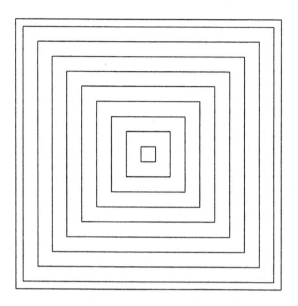

FIG. 1.6.1 - *Contour lines of the function U in* (1.6.2).

FIG. 1.6.2 - *Contour lines of $q_{\epsilon,n}$ for $n = 16$ and $\epsilon = \frac{1}{20}$.*

FIG. 1.6.3 - *Contour lines of $q_{\epsilon,n}$ for $n = 16$ and $\epsilon = \frac{1}{40}$.*

FIG. 1.6.4 - *Contour lines of $q_{\epsilon,n}$ for $n = 20$ and $\epsilon = \frac{1}{40}$.*

Since U_ϵ, which is regular and concave, converges for $\epsilon \to 0$ to the non-regular function U in (1.6.2), when ϵ is small meaningful approximated solutions are only obtained for very large n. This behavior has no relation to that, due to the influence of transport terms, affecting the equations which will be examined in the following chapters.

Piece-wise linear finite elements, on a regular triangular mesh, would be probably more appropriate for the approximation of this problem. The purpose, however, was to demonstrate that spectral methods can also be employed in the case of non-regular solutions. A 3-D version of the eikonal equation will be considered in section 6.4.

1.7 More about boundary conditions

In this section, we introduce a different notation for the matrix and the right-hand side vector of (1.2.6). We begin by defining two singular $(n+1)^2 \times (n+1)^2$ matrices: $\mathcal{J}_n^{\partial\Omega}$ and $\mathcal{J}_n^{\Omega}$. The multiplication of $\mathcal{J}_n^{\partial\Omega}$ by a vector $\vec{X}_n$ sets to zero the $(n-1)^2$ entries corresponding to the Legendre nodes inside Ω without changing the other entries. The multiplication of $\mathcal{J}_n^{\Omega}$ by $\vec{X}_n$ sets to zero the $4n$ entries corresponding to the boundary nodes without changing the other entries.

It is clear that

$$(1.7.1) \qquad \mathcal{J}_n^{\Omega} + \mathcal{J}_n^{\partial\Omega} = \mathcal{I}_n \qquad \text{and} \qquad \mathcal{J}_n^{\Omega} \mathcal{J}_n^{\partial\Omega} = 0,$$

where $\mathcal{I}_n$ is the identity matrix.

Then, we define $\vec{F}_n$ to be the vector of the values $f(\Theta_k^{(n)})$, $0 \le k \le n_T$ (see (1.2.3)). Similarly, $\vec{G}_n$ will be the vector containing the values of the functions g_k, $1 \le k \le 4$, at the entries associated with the boundary points, while the other entries are required to be zero. For example, in the case $n = 4$ one has

$$\vec{G}_4 \equiv \left(g_1(\eta_0^{(n)}),\ g_1(\eta_1^{(n)}),\ g_1(\eta_2^{(n)}),\ g_1(\eta_3^{(n)}),\ g_2(\eta_0^{(n)}), \right.$$

$$g_4(\eta_1^{(n)}),\quad 0\quad,\quad 0\quad,\quad 0\quad,\ g_2(\eta_1^{(n)}),$$

$$g_4(\eta_2^{(n)}),\quad 0\quad,\quad 0\quad,\quad 0\quad,\ g_2(\eta_2^{(n)}),$$

$$g_4(\eta_3^{(n)}),\quad 0\quad,\quad 0\quad,\quad 0\quad,\ g_2(\eta_3^{(n)}),$$

$$\left. g_4(\eta_4^{(n)}),\ g_3(\eta_1^{(n)}),\ g_3(\eta_2^{(n)}),\ g_3(\eta_3^{(n)}),\ g_3(\eta_4^{(n)}) \right).$$

We note that $\vec{G}_n = \mathcal{J}_n^{\partial\Omega} \vec{G}_n$ and that $A_n = \mathcal{J}_n^{\Omega} \mathcal{L}_n + \mathcal{J}_n^{\partial\Omega}$, where $\mathcal{L}_n$ is the discretization of the operator $-\Delta$ in the space $\mathbf{P}_n^*$.

Therefore, the system (1.2.6) can be written in equivalent form as

$$(1.7.2) \qquad (\mathcal{J}_n^{\Omega} \mathcal{L}_n + \mathcal{J}_n^{\partial\Omega}) \vec{X}_n = \mathcal{J}_n^{\Omega} \vec{F}_n + \vec{G}_n,$$

which by virtue of (1.7.1) is also equivalent to

$$(1.7.3) \qquad \mathcal{J}_n^{\Omega} \mathcal{L}_n \vec{X}_n = \mathcal{J}_n^{\Omega} \vec{F}_n \quad \text{and} \quad \mathcal{J}_n^{\partial\Omega} \vec{X}_n = \vec{G}_n.$$

1.8   A Sturm-Liouville problem

We analyze the approximation of an elliptic differential equation that will be useful later in sections 5.2 and 5.3. Let Ω be the square $]-1,1[\times]-1,1[$, and let $b : \bar{\Omega} \to \mathbf{R}$ be the function defined by $b(x,y) := (1-x^2)(1-y^2)$, $(x,y) \in \bar{\Omega}$. Then, we look for the solution $\mathcal{P} : \Omega \to \mathbf{R}$ to the equation

$$(1.8.1) \quad -\text{div}(b \, \vec{\nabla}\mathcal{P}) = -\frac{\partial}{\partial x}\left(b \, \frac{\partial \mathcal{P}}{\partial x}\right) - \frac{\partial}{\partial y}\left(b \, \frac{\partial \mathcal{P}}{\partial y}\right) = f \qquad \text{in } \Omega,$$

where $f : \Omega \to \mathbf{R}$ is a given right-hand side function. No boundary conditions will be imposed, but we shall assume the additional constraint

$$(1.8.2) \qquad \int_{\Omega} \mathcal{P} \, dx dy = 0.$$

In (1.8.1) we have a *Sturm-Liouville* problem of *singular* type, since b is vanishing on $\partial\Omega$. Integrating the terms of (1.8.1) in Ω, we discover that f must satisfy the compatibility condition

$$(1.8.3) \qquad \int_{\Omega} f \, dx dy = 0.$$

Following (1.1.6), we write our equation in variational form

$$(1.8.4) \qquad B(\mathcal{P}, \phi) := \int_{\Omega} b \left(\frac{\partial \mathcal{P}}{\partial x} \frac{\partial \phi}{\partial x} + \frac{\partial \mathcal{P}}{\partial y} \frac{\partial \phi}{\partial y} \right) dx dy$$

$$= \int_{\Omega} f \phi \, dx dy =: F(\phi),$$

and we soon recognize, thanks to (5.2.8), that the bilinear form B is strictly positive definite (i.e. satisfies the second inequality in (1.1.3)). Applying theorem 1.1.2, one should be able to prove the existence and the uniqueness of a solution satisfying (1.8.2) and (1.8.4). Such a solution will belong to a suitable functional space **X**.

The condition (1.8.3) is necessary for existence, while the condition (1.8.2) is sufficient for uniqueness (note that, for $f = 0$, all the constant functions $\mathcal{P} = c \in \mathbf{R}$ satisfy (1.8.1), but the only constant realizing (1.8.2) is $c = 0$).

Concerning the numerical approximation, we take into account the set of Gauss nodes $\xi_k^{(n)}$, $1 \leq k \leq n$, i.e. the n zeroes of the Legendre polynomial P_n (see section $A.2$). Since boundary conditions are not required, this time, the approximated solution $r_n \in \mathbf{P}_{n-1}^*$ will be obtained by collocating (1.8.1) at the Gauss nodes, i.e.

$$(1.8.5) \qquad -\left[\frac{\partial}{\partial x}\left(b\,\frac{\partial r_n}{\partial x}\right) + \frac{\partial}{\partial y}\left(b\,\frac{\partial r_n}{\partial y}\right)\right](\xi_j^{(n)},\xi_i^{(n)}) = f(\xi_j^{(n)},\xi_i^{(n)}),$$

for $1 \leq i \leq n$ and $1 \leq j \leq n$. With the help of the quadrature formula $(A.2.1)$, we get (1.8.2) by imposing

$$(1.8.6) \qquad \int_\Omega r_n\,dx\,dy = \sum_{i,j=1}^n r_n(\xi_j^{(n)},\xi_i^{(n)})\,w_i^{(n)}w_j^{(n)} = 0.$$

We now rewrite (1.8.5) as a linear system. In order to compute the entries of the matrix, we introduce the set of Lagrange polynomials

$$(1.8.7) \qquad l_m^{(n)}(x) := \frac{P_n(x)}{P_n'(\xi_m^{(n)})}\frac{1}{x - \xi_m^{(n)}} \qquad x \neq \xi_m^{(n)}, \quad 1 \leq m \leq n,$$

with $l_m^{(n)}(\xi_m^{(n)}) := 1$. We have $l_m^{(n)} \in \mathbf{P}_{n-1}$, $1 \leq m \leq n$ and $l_m^{(n)}(\xi_j^{(n)}) = 0$ for $m \neq j$. An easy computation, carried out with the help of $(A.1.5)$, shows that

$$(1.8.8) \qquad \hat{d}_{jm} := -\left[\frac{d}{dx}\left((1-x^2)\frac{d}{dx}l_m^{(n)}\right)\right](\xi_j^{(n)})$$

$$= \begin{cases} \dfrac{2\,(1 - [\xi_j^{(n)}]^2)}{(\xi_j^{(n)} - \xi_m^{(n)})^2}\dfrac{P_n'(\xi_j^{(n)})}{P_n'(\xi_m^{(n)})} & \text{if } j \neq m, \\[4ex] \frac{1}{3}\left[n(n+1) - 2(1 - [\xi_m^{(n)}]^2)^{-1}\right] & \text{if } j = m. \end{cases}$$

To get the coefficients in (1.8.8), one can follow the path outlined in GOTTLIEB, HUSSAINI and ORSZAG (1984) with the aim of obtaining the entries of other matrices relevant to the theory of spectral methods.

Thus, an expression similar to (1.2.4) is

(1.8.9) $$-\text{div}\big(b\,\vec{\nabla}r_n\big)(\xi_j^{(n)},\xi_i^{(n)})$$

$$= \sum_{m=1}^{n} \big[(1 - [\xi_i^{(n)}]^2)\hat{d}_{jm}\; r_n(\xi_m^{(n)},\xi_i^{(n)}) + (1 - [\xi_j^{(n)}]^2)\hat{d}_{im}\; r_n(\xi_j^{(n)},\xi_m^{(n)})\big],$$

for $1 \le i \le n$, $1 \le j \le n$. We find out that our matrix of dimension n^2, for $n = 5$, has the same structure as that of Fig. 1.2.2. In this way (1.8.5) and (1.8.6) lead to a set of $n^2 + 1$ linear conditions. We note instead that a polynomial in $\mathbf{P}_{n-1}^{\star}$ depends on n^2 degrees of freedom. Hence, we need to get rid of an extra equation in (1.8.5) which is linearly dependent on the others. This will be automatically done by solving iteratively the linear system corresponding to (1.8.5). We apply the Du Fort-Frankel method (1.4.6) to the system (1.2.6) where now $\mathcal{A}_n$ denotes the matrix associated with (1.8.9), and the vectors $\vec{X}_n$ and $\vec{B}_n$ of length n^2 contain respectively the values of r_n and f at the collocation nodes. At each step $k \ge 1$ of the iterative method, we enforce the condition (1.8.6) by correcting the entries of $\vec{X}_n^k$. Actually, $\vec{X}_n^k$ represents the values of a polynomial $r_n^{(k)} \in \mathbf{P}_{n-1}^{\star}$ at the collocation nodes, so that, if $r_n^{(k)}$ does not have a zero average as (1.8.6) requires, we will replace it by $r_n^{(k)} - c_k$ where $c_k := \frac{1}{4}\int_{\Omega} r_n^{(k)}\,dx dy$.

E	$n = 4$	$n = 8$	$n = 12$	$n = 16$	$n = 20$
10^{-1}	6	27	63	112	175
10^{-2}	11	58	135	242	381
10^{-3}	16	88	207	373	586
10^{-4}	21	118	279	503	792
10^{-5}	26	148	350	634	997
10^{-6}	31	178	422	764	1203

TABLE 1.8.1 - *Iterations needed to obtain the norm of the residual less than E.*

As preconditioning matrix $\mathcal{B}_n$ we take the diagonal of $\mathcal{A}_n$, i.e. the vector $\{(1-[\xi_i^{(n)}]^2)\hat{d}_{jj}+(1-[\xi_j^{(n)}]^2)\hat{d}_{ii}\}_{1\le i\le n,1\le j\le n}$. The preconditioned eigenvalues depend on n which results in a mild slow down of the convergence for larger n.

Nevertheless, considering that the cost of each iteration is only proportional to n^3 and that there is no pre-processing to factorize the preconditioner, the computational time to carry out the approximated solution up to a certain accuracy is still competitive with respect to the time needed by a direct solver. Therefore, we will not try to enlarge the band of $\mathcal{B}_n$ in order to improve the convergence rate of the Du Fort-Frankel method, although we guess that the performances of this solution technique could be considerably ameliorated.

A good convergence rate is achieved when the parameters in (1.4.7) are respectively $\sigma_1 = .75$ and $\sigma_2 = .9$, even though this may not be the optimal choice for any n.

To show the results of a numerical test, we take $f(x,y) = \sin xy$, $(x,y) \in \Omega$, which satisfies (1.8.3). In Table 1.8.1, for different values of n, we report the number of iterations in order to obtain that the norm of the residual

$$(1.8.10) \qquad \|A_n \vec{X}_n^k - \vec{B}_n\|$$

$$:= \left(\sum_{i,j=1}^{n} \left[-\operatorname{div}(b\vec{\nabla}r_n^{(k)}) - f \right]^2 (\xi_j^{(n)}, \xi_i^{(n)}) \, w_i^{(n)} w_j^{(n)} \right)^{1/2}$$

is less than a prescribed accuracy E. The weights of the quadrature formula (1.8.10) are defined in $(A.2.2)$. The initial vectors $\vec{X}_n^0$ and $\vec{X}_n^1$ in this experiment were set to zero. A reduction of the number of iterations may be obtained by better adjusting the parameters σ_1 and σ_2.

2

Steady Transport-Diffusion Equations

To more general elliptic boundary-value problems we apply the numerical technique introduced in the previous chapter. Namely, we are interested in differential operators in which the second-order terms are dominated by the first-order terms, giving rise to the boundary layer phenomenon. To better deal with these equations, a modified version of the Legendre collocation grid will be considered.

2.1 A more general boundary-value problem

In the square $\Omega =]-1,1[\times]-1,1[$, we look for the solution $U : \bar{\Omega} \to \mathbf{R}$ to the equation

$$(2.1.1) \quad f_1 \frac{\partial^2 U}{\partial x^2} + f_2 \frac{\partial^2 U}{\partial x \partial y} + f_3 \frac{\partial^2 U}{\partial y^2} + f_4 \frac{\partial U}{\partial x} + f_5 \frac{\partial U}{\partial y} + f_6 U = f_7,$$

where f_k, $1 \leq k \leq 7$, are given functions defined in $\bar{\Omega}$. In addition, we require that U satisfies the same Dirichlet boundary conditions provided in (1.1.2). We assume that the differential operator, written in *non-conservative* form:

$$(2.1.2) \quad L := f_1 \frac{\partial^2}{\partial x^2} + f_2 \frac{\partial^2}{\partial x \partial y} + f_3 \frac{\partial^2}{\partial y^2} + f_4 \frac{\partial}{\partial x} + f_5 \frac{\partial}{\partial y} + f_6,$$

is of *elliptic* type (see SOBOLEV (1964) or WEINBERGER (1965)), i.e. that the coefficients f_k, $1 \leq k \leq 3$ are such that

$$(2.1.3) \quad f_1 a^2 + f_2 ab + f_3 b^2 < 0 \quad \text{in } \Omega, \quad \forall (a,b) \in \mathbf{R}^2, \ (a,b) \neq (0,0).$$

We do not undertake any theoretical analysis related to the problem. From now on, existence, uniqueness and regularity of U will be taken for granted.

Following the terminology adopted for instance in fluid dynamics, we will refer to the second-order terms of the differential operator L in (2.1.2) as the *diffusive* part, and the first-order terms of L as the *advective* or *transport* part. The choice $f_1 = f_3 = -1$, $f_2 = f_4 = f_5 = f_6 = 0$, brings us back to the Poisson equation already studied in the previous chapter. Here, we shall mainly focus our attention on the cases in which the diffusion terms are much smaller than the transport terms. The case corresponding to $f_1 = f_3 = -\epsilon$, $f_2 = f_6 = 0$, $f_4 = \beta_1$, $f_5 = \beta_2$, where the small parameter $\epsilon > 0$ and the vector $\vec{\beta} = (\beta_1, \beta_2) \in \mathbf{R}^2$ are given, leads to the equation

$$(2.1.4) \qquad\qquad -\epsilon \, \Delta U + \vec{\beta} \cdot \vec{\nabla} U = f_7 \quad \text{in } \Omega,$$

which is a typical example of a transport-dominated equation that will be extensively examined in the coming sections. Actually, (2.1.4) is the steady version of the advection-diffusion equation (5.1.1), which is the kernel of many mathematical models. If we assume homogeneous boundary conditions (i.e., $g_k = 0$, $1 \leq k \leq 4$), it is easy to prove that the solution to (2.1.4) satisfies (1.1.7) with f_7 in place of f.

We define the *flux* to be the vector $\epsilon^{-1}\vec{\beta}$. For ϵ decreasing, the solution U moves in the direction of the flux, and generates the so-called *boundary layers* in the neighborhood of the points of $\partial\Omega$ where $\vec{\beta} \cdot \vec{\nu} > 0$, $\vec{\nu}$ being the outward normal vector to the boundary of Ω. It is well-known that the numerical treatment of solutions presenting boundary layers is not a trivial matter, and many *stabilization* techniques have been developed within the framework of finite-differences or finite elements (see QUARTERONI and VALLI (1994), section 8.3), with the purpose of obtaining decent approximations at low cost. Classical references in this field are BROOKS and HUGHES (1982), JOHNSON (1987), DOUGLAS and WANG (1989), HUGHES, FRANCA and HULBERT (1989). The more recent paper of BAIOCCHI, BREZZI and FRANCA (1993), introducing the concept of *virtual bubbles*, establishes a coherent unifying theory for various classes of stabilization methods available for advection-dominated equations. We remind the reader that a stabilized approximated solution does not explode when $\epsilon \to 0$ and the discretization parameter is fixed.

Of course, difficulties also arise for spectral methods. We examine the numerical approximation of (2.1.1)-(1.1.2) by the collocation method, which means looking for $q_n \in \mathbf{P}_n^\star$ such that

$$(2.1.5) \qquad\qquad (Lq_n)(\eta_j^{(n)}, \eta_i^{(n)}) = f_7(\eta_j^{(n)}, \eta_i^{(n)}),$$

for $1 \leq i \leq n-1$, $1 \leq j \leq n-1$. Moreover, we impose the set of boundary conditions in (1.2.2). One may develop a convergence analysis of q_n to U

for $n \to +\infty$, assuming the regularity of the coefficients f_k, $1 \leq k \leq 7$. Nevertheless, the idea of collocating the equation (2.1.1) at the Legendre grid, which was very effective in the case of the Poisson equation, does not seem to be the right choice in the case of other differential problems, such as (2.1.4) for small ϵ. A good resolution of the boundary layers is only recovered when n is large. Usually, the estimate $n > c/\sqrt{\epsilon}$, where $c > 0$ is a suitable constant, tells us that the degree n should be adequate (see CANUTO (1988)). For ϵ small and a degree n not sufficiently large to resolve the layer ($n \ll c/\sqrt{\epsilon}$), the collocation method at the Legendre nodes leads to approximated solutions which are polluted by high frequency oscillations. For n tending to infinity, the oscillations disappear and the approximated solutions smoothly converge to the exact one, as predicted by the theory. On the other hand, for n fixed and ϵ tending to zero, the amplitude of these spurious oscillations tends to infinity. Unfortunately, many applications require ϵ to be so small that an appropriate choice of n, aimed to avoid unpleasant oscillatory phenomena, is too costly.

We need to construct a new collocation grid, as we are going to explain in the next section. The idea that we shall follow is quite general and can be applied in other circumstances. We enlarge $\mathbf{P}_n^*$ by adding the space spanned by a polynomial $\chi_n \in \mathbf{P}_{n+1}^{*,0}$. Then, we look for the new collocation nodes among the zeros of the function $L\chi_n$. Since χ_n is such that the Legendre nodes are the zeros of $L\chi_n$ when $L = -\Delta$ is the Laplace operator, this approach turns out to be an extension of the usual collocation method presented in chapter one. The collocation method based on the new grid will bring an evident reduction of the oscillations. With the new approach, for n fixed and ϵ tending to zero, the corresponding approximated solutions converge to a limit, which is a polynomial representing a rough but consistent idea of the behavior of the exact solution. This will allow to generate reliable approximations near and far from the boundary layers, also in the case in which the degree n is not high enough.

2.2   The upwind grid

We introduce a new set of grid points in $\bar{\Omega}$, which is no longer in the form of a Cartesian product. The grid points are denoted by $(\tau_{i,j}^{(n)}, \upsilon_{i,j}^{(n)})$, $0 \leq i \leq n$, $0 \leq j \leq n$, and coincide with the usual Legendre nodes $(\eta_j^{(n)}, \eta_i^{(n)})$ at the boundary $\partial\Omega$, so that

$$(2.2.1) \quad \begin{cases} \tau_{0,j}^{(n)} = \tau_{n,j}^{(n)} := \eta_j^{(n)} \quad & \upsilon_{0,j}^{(n)} := -1 \quad & \upsilon_{n,j}^{(n)} := 1 \quad & 0 \leq j \leq n, \\ \upsilon_{i,0}^{(n)} = \upsilon_{i,n}^{(n)} := \eta_i^{(n)} \quad & \tau_{i,0}^{(n)} := -1 \quad & \tau_{i,n}^{(n)} := 1 \quad & 0 \leq i \leq n. \end{cases}$$

Inside Ω, i.e. for $1 \le i \le n-1$, $1 \le j \le n-1$, each $(\tau_{i,j}^{(n)}, v_{i,j}^{(n)})$ is a suitable perturbation of $(\eta_j^{(n)}, \eta_i^{(n)})$ determined according to the size of the coefficients f_k, $1 \le k \le 7$, of the differential operator L in (2.1.2). In order to get the coordinates of the new nodes, we fix i and j and compute x^* and y^* to be the zeroes of a linear combination of the polynomials P_n and P_n':

$$(2.2.2) \qquad \Upsilon_0\, P_n'(x^*) - \Upsilon_1\, P_n(x^*) = 0,$$

with $\Upsilon_1 := (1 - [\eta_i^{(n)}]^2)\beta_1$ and

$$\beta_1 := f_4(\eta_j^{(n)}, \eta_i^{(n)}) - \frac{2\eta_i^{(n)}}{1 - [\eta_i^{(n)}]^2} f_2(\eta_j^{(n)}, \eta_i^{(n)}),$$

$$(2.2.3) \qquad \Upsilon_0\, P_n'(y^*) - \Upsilon_2\, P_n(y^*) = 0,$$

with $\Upsilon_2 := (1 - [\eta_j^{(n)}]^2)\beta_2$ and

$$\beta_2 := f_5(\eta_j^{(n)}, \eta_i^{(n)}) - \frac{2\eta_j^{(n)}}{1 - [\eta_j^{(n)}]^2} f_2(\eta_j^{(n)}, \eta_i^{(n)}),$$

where we also defined

$$\Upsilon_0 := \Big[- f_1(\eta_j^{(n)}, \eta_i^{(n)})(1 - [\eta_i^{(n)}]^2) - f_3(\eta_j^{(n)}, \eta_i^{(n)})(1 - [\eta_j^{(n)}]^2)$$
$$+ \frac{1}{n(n+1)}\, f_6(\eta_j^{(n)}, \eta_i^{(n)})(1 - [\eta_i^{(n)}]^2)(1 - [\eta_j^{(n)}]^2) \Big].$$

More precisely, each one of the two equations (2.2.2) and (2.2.3) admits n zeroes (see FUNARO (1997)). We do not need to compute all of them, as we are only concerned with the couple (x^*, y^*) which is nearest to the point $(\eta_j^{(n)}, \eta_i^{(n)})$. We assume that $\Upsilon_0 > 0$, which is in agreement with the ellipticity of the operator L expressed by the condition (2.1.3).

We note that $x^* = \eta_j^{(n)} \Leftrightarrow \beta_1 = 0$ and $y^* = \eta_i^{(n)} \Leftrightarrow \beta_2 = 0$. Then, we define

$$(2.2.4) \qquad \rho := \begin{cases} \Big[\beta_1^2(x^* - \eta_j^{(n)})^{-2} \\ \quad + \beta_2^2(y^* - \eta_i^{(n)})^{-2}\Big]^{-\frac{1}{2}} & \text{if } x^* \ne \eta_j^{(n)},\ y^* \ne \eta_i^{(n)}, \\[2mm] |\beta_1|^{-1}|x^* - \eta_j^{(n)}| & \text{if } x^* \ne \eta_j^{(n)},\ y^* = \eta_i^{(n)}, \\[2mm] |\beta_2|^{-1}|y^* - \eta_i^{(n)}| & \text{if } x^* = \eta_j^{(n)},\ y^* \ne \eta_i^{(n)}, \\[2mm] 0 & \text{if } x^* = \eta_j^{(n)},\ y^* = \eta_i^{(n)}. \end{cases}$$

Finally, we set

(2.2.5) $$\tau_{i,j}^{(n)} := \eta_j^{(n)} - \rho\beta_1, \qquad \upsilon_{i,j}^{(n)} := \eta_i^{(n)} - \rho\beta_2.$$

We explain the reasons for this construction. The equations in (2.2.5) say that the point $(\tau_{i,j}^{(n)}, \upsilon_{i,j}^{(n)})$ lies on the straight line of equation $\beta_2(x - \eta_j^{(n)}) = \beta_1(y - \eta_i^{(n)})$, in the direction opposite to the one of vector $\vec{\beta} := (\beta_1, \beta_2)$. On the other hand, because of (2.2.4), the point $(\tau_{i,j}^{(n)}, \upsilon_{i,j}^{(n)})$ also belongs to the ellipse of equation

(2.2.6) $$\left(\frac{x - \eta_j^{(n)}}{x^* - \eta_j^{(n)}} \right)^2 + \left(\frac{y - \eta_i^{(n)}}{y^* - \eta_i^{(n)}} \right)^2 = 1.$$

The ellipse can possibly degenerate into a segment.

Introducing the zeroes $\xi_i^{(n)} \in]-1, 1[$, $1 \le i \le n$, of the Legendre polynomial P_n (see section A.2), it is not difficult to realize that

(2.2.7) $$\xi_j^{(n)} < x^* < \xi_{j+1}^{(n)} \quad \text{and} \quad \xi_i^{(n)} < y^* < \xi_{i+1}^{(n)}.$$

Therefore, the point $(\tau_{i,j}^{(n)}, \upsilon_{i,j}^{(n)})$ falls inside the rectangle $]\xi_j^{(n)}, \xi_{j+1}^{(n)}[\times]\xi_i^{(n)}, \xi_{i+1}^{(n)}[$. This situation is illustrated in Fig. 2.2.1. Moreover, if the ratio $|\Upsilon_1/\Upsilon_0|$ is large, then the point x^* approaches to $\xi_j^{(n)}$ if $\beta_1 > 0$, while it approaches to $\xi_{j+1}^{(n)}$ if $\beta_1 < 0$. A similar behavior is exhibited by y^* when the ratio $|\Upsilon_2/\Upsilon_0|$ is large.

This new set of nodes is called the *upwind grid* and was first introduced in FUNARO (1993). We immediately observe that in the case $f_2 = f_4 = f_5 = 0$, one has $\tau_{i,j}^{(n)} = \eta_j^{(n)}$ and $\upsilon_{i,j}^{(n)} = \eta_i^{(n)}$, for $0 \le i \le n$, $0 \le j \le n$. Therefore, in the case of the Poisson equation, the upwind grid coincides with the standard Legendre grid.

The roots x^* and y^* in (2.2.2) and (2.2.3) can be determined for example with the help of the *secant method* (a few steps are usually sufficient). An approximate value of x^* is given by the expressions

(2.2.8) $$\begin{cases} \dfrac{\Upsilon_0 n(n+1)\eta_j^{(n)} P_n(\eta_j^{(n)}) + \Upsilon_1 \xi_j^{(n)} P'(\xi_j^{(n)})(1 - [\eta_j^{(n)}]^2)}{\Upsilon_0 n(n+1) P_n(\eta_j^{(n)}) + \Upsilon_1 P'(\xi_j^{(n)})(1 - [\eta_j^{(n)}]^2)} & \beta_1 \ge 0 \\[4ex] \dfrac{\Upsilon_0 n(n+1)\eta_j^{(n)} P_n(\eta_j^{(n)}) + \Upsilon_1 \xi_{j+1}^{(n)} P'(\xi_{j+1}^{(n)})(1 - [\eta_j^{(n)}]^2)}{\Upsilon_0 n(n+1) P_n(\eta_j^{(n)}) + \Upsilon_1 P'(\xi_{j+1}^{(n)})(1 - [\eta_j^{(n)}]^2)} & \beta_1 \le 0, \end{cases}$$

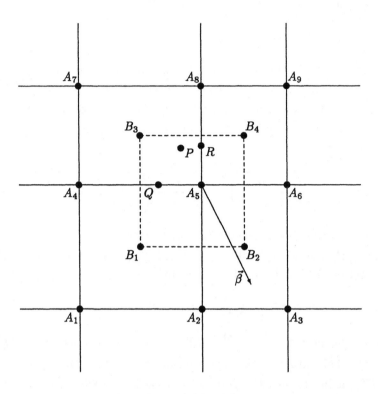

FIG. 2.2.1 - *Position of the point $P \equiv (\tau_{i,j}^{(n)}, v_{i,j}^{(n)})$ in relation to the Legendre grid points. We have set:* $A_1 \equiv (\eta_{j-1}^{(n)}, \eta_{i-1}^{(n)})$, $A_2 \equiv (\eta_{j}^{(n)}, \eta_{i-1}^{(n)})$, $A_3 \equiv (\eta_{j+1}^{(n)}, \eta_{i-1}^{(n)})$, $A_4 \equiv (\eta_{j-1}^{(n)}, \eta_{i}^{(n)})$, $A_5 \equiv (\eta_{j}^{(n)}, \eta_{i}^{(n)})$, $A_6 \equiv (\eta_{j+1}^{(n)}, \eta_{i}^{(n)})$, $A_7 \equiv (\eta_{j-1}^{(n)}, \eta_{i+1}^{(n)})$, $A_8 \equiv (\eta_{j}^{(n)}, \eta_{i+1}^{(n)})$, $A_9 \equiv (\eta_{j+1}^{(n)}, \eta_{i+1}^{(n)})$, $B_1 \equiv (\xi_{j}^{(n)}, \xi_{i}^{(n)})$, $B_2 \equiv (\xi_{j+1}^{(n)}, \xi_{i}^{(n)})$, $B_3 \equiv (\xi_{j}^{(n)}, \xi_{i+1}^{(n)})$, $B_4 \equiv (\xi_{j+1}^{(n)}, \xi_{i+1}^{(n)})$, $Q \equiv (x^*, \eta_{i}^{(n)})$, $R \equiv (\eta_{j}^{(n)}, y^*)$.

and an approximate value of y^* is given by the expressions

$$(2.2.9) \quad \begin{cases} \dfrac{\Upsilon_0 n(n+1)\eta_i^{(n)} P_n(\eta_i^{(n)}) + \Upsilon_2 \xi_i^{(n)} P'(\xi_i^{(n)})(1 - [\eta_i^{(n)}]^2)}{\Upsilon_0 n(n+1) P_n(\eta_i^{(n)}) + \Upsilon_2 P'(\xi_i^{(n)})(1 - [\eta_i^{(n)}]^2)} & \beta_2 \geq 0 \\[4mm] \dfrac{\Upsilon_0 n(n+1)\eta_i^{(n)} P_n(\eta_i^{(n)}) + \Upsilon_2 \xi_{i+1}^{(n)} P'(\xi_{i+1}^{(n)})(1 - [\eta_i^{(n)}]^2)}{\Upsilon_0 n(n+1) P_n(\eta_i^{(n)}) + \Upsilon_2 P'(\xi_{i+1}^{(n)})(1 - [\eta_i^{(n)}]^2)} & \beta_2 \leq 0. \end{cases}$$

Using the recursion formula for the Legendre polynomials $(A.1.6)$ and five or six steps of the secant method, the cost of determining x^* and y^* is proportional to n. Thus, the cost of computing all the nodes of the upwind grid is proportional to n^3.

As anticipated at the end of section 2.1, the construction of the upwind grid was made by keeping in mind, for $1 \leq i \leq n - 1$, $1 \leq j \leq n - 1$, the determination of the point (x^*, y^*) satisfying the relations

$$(2.2.10) \quad \begin{cases} (L\chi_n)(x^*, y^*) = 0, \\ \beta_2(x^* - \eta_j^{(n)}) = \beta_1(y^* - \eta_i^{(n)}), \end{cases}$$

where L is the operator in $(2.1.2)$ and the polynomial $\chi_n \in \mathbf{P}_{n+1}^{*,0}$ is given by $\chi_n(x, y) := (1 - x^2)(1 - y^2)P_n'(x)P_n'(y)$, $(x, y) \in \bar{\Omega}$.

The set of nodes generated with the help of the formulas $(2.2.2)$ and $(2.2.3)$ do not satisfy exactly $(2.2.10)$, with the exception of some special cases. As a matter of fact, the node $(\tau_{i,j}^{(n)}, v_{i,j}^{(n)})$ is near $(\eta_j^{(n)}, \eta_i^{(n)})$ in the direction contrary to $\vec{\beta}$, but $(L\chi_n)(\tau_{i,j}^{(n)}, v_{i,j}^{(n)})$ is not zero. This node is very close to (x^*, y^*), however, and the error committed is not so crucial for the applications that follow.

We check that $(2.2.10)$ is true for a particular case. We assume that f_1 and f_4 are constant, and $f_2 = f_3 = f_5 = f_6 = 0$, although this choice would not be allowed by condition $(2.1.3)$. Hence, the vector $\vec{\beta}$ now takes the form $(f_4, 0)$. Finally, we apply the operator L to the polynomial χ_n, and using the differential equation $((1 - x^2)P_n'(x))' = -n(n + 1)P_n(x)$, $x \in [-1, 1]$ (see $(A.1.5)$), we get

$$(2.2.11) \quad \left[f_1 \frac{\partial^2 \chi_n}{\partial x^2} + f_4 \frac{\partial \chi_n}{\partial x} \right](x, y)$$

$$= f_1 \left[-n(n + 1)P_n'(x)(1 - y^2)P_n'(y) \right]$$

$$+ f_4 \left[-n(n + 1)P_n(x)(1 - y^2)P_n'(y) \right]$$

$$= \left[\Upsilon_0 P_n'(x) - \Upsilon_1 P_n(x) \right] \frac{n(n + 1)(1 - y^2)}{1 - [\eta_i^{(n)}]^2} P_n'(y).$$

By setting to zero the last term in $(2.2.11)$, we get a characterization of the upwind nodes of the type $(x^*, \eta_i^{(n)})$, $1 \leq i \leq n - 1$. In fact, we have $(L\chi_n)(x^*, \eta_i^{(n)}) = 0$. As we already noticed, one also has $x^* \in]\xi_j^{(n)}, \eta_j^{(n)}[$ if $f_4 > 0$, $x^* \in]\eta_j^{(n)}, \xi_{j+1}^{(n)}[$ if $f_4 < 0$, and $x^* = \eta_j^{(n)}$ if $f_4 = 0$. Further explanations are given in FUNARO (1993).

The upwind grid takes into account the magnitude and the sign of the transport terms: if the transport part of L strongly dominates the diffusive part, the upwind grid is far away from the Legendre grid, staying however inside the limits established in Fig. 2.2.1.

One may guess that

$$(2.2.12) \qquad \lim_{n \to +\infty} (\tau_{i,j}^{(n)}, \upsilon_{i,j}^{(n)}) = (\eta_j^{(n)}, \eta_i^{(n)}), \qquad 0 \le i \le n, \ 0 \le j \le n,$$

where the convergence rate is of the order of n^{-2} at the center of the square Ω and n^{-4} near the boundary. This implies that the diameter of the ellipse (2.2.6) tends to zero faster than the diameter of the square $B_1 B_2 B_3 B_4$ of Fig. 2.2.1.

Finally, we note that the upwind grid can also be defined when the coefficients of the second-order terms of L are vanishing, a fact that will be used in section 6.2 for the approximation of first-order equations. Taking for instance $\epsilon = 0$ in (2.1.4), we have $\Upsilon_0 = 0$, so that from (2.2.2) we get $x^* = \xi_j^{(n)}$ or $x^* = \xi_{j+1}^{(n)}$, depending on the sign of β_1, and from (2.2.3) we get $y^* = \xi_i^{(n)}$ or $y^* = \xi_{i+1}^{(n)}$, depending on the sign of β_2.

Following are some examples of grids. Remember that the usual Legendre grid is given in Fig. 1.2.1 for $n = 10$. Concerning the differential operator $-\epsilon\Delta + \vec{\beta} \cdot \vec{\nabla}$ (see equation (2.1.4)), when $\epsilon = .05$ and $n = 10$, we find in Figs. 2.2.2, 2.2.3, 2.2.4 the upwind grids corresponding to the vectors $\vec{\beta} = (1,0)$, $\vec{\beta} = (0,2)$, $\vec{\beta} = (1,2)$, respectively. In Fig. 2.2.5, always for $n = 10$, we considered the case of the operator $L = -\frac{\partial^2}{\partial x^2} + 2\frac{\partial^2}{\partial x \partial y} - \frac{\partial^2}{\partial y^2}$ (note that L does not satisfy (2.1.3)). Further grids related to the operator $-\epsilon\Delta + \vec{\beta} \cdot \vec{\nabla}$, with $\vec{\beta}$ depending on x and y, are presented for $n = 10$ in Figs. 2.2.6 and 2.2.7.

———— ◇ ————

It will be useful later on to pass from the values of a polynomial $r_n \in \mathbf{P}_n^*$, given at the Legendre grid, to its values at the upwind grid. This operation can be done by exact interpolation by using, independently for the x and y directions, the formula (A.3.3). For example, we can determine the value of r_n at a point $(\hat{x}, \hat{y})$ by first computing the values $r_n(\hat{x}, \eta_i^{(n)})$, $0 \le i \le n$, with the help of (A.3.3) applied $n+1$ times in the x direction. Then, we apply one more time (A.3.3) in the y direction to compute $r_n(\hat{x}, \hat{y})$. Therefore, it turns out that the cost of the global two-dimensional interpolation is proportional to n^4, and cannot be reduced since one realizes that the grid $(\tau_{i,j}^{(n)}, \upsilon_{i,j}^{(n)})$, $0 \le i \le n$, $0 \le j \le n$, is very unstructured. This price is too high for applications, so we propose an alternative interpolation algorithm.

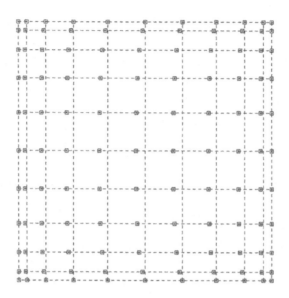

FIG. 2.2.2 - *The upwind grid corresponding to* $L = -\frac{1}{20}\Delta + \frac{\partial}{\partial x}$ *for* $n = 10$.

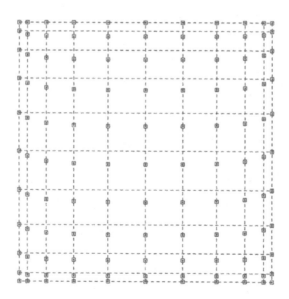

FIG. 2.2.3 - *The upwind grid corresponding to* $L = -\frac{1}{20}\Delta + 2\frac{\partial}{\partial y}$ *for* $n = 10$.

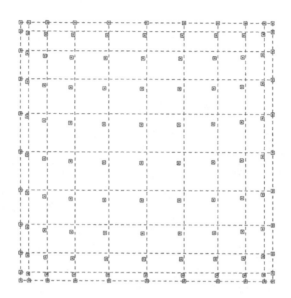

FIG. 2.2.4 - *The upwind grid corresponding to* $L = -\frac{1}{20}\Delta + \frac{\partial}{\partial x} + 2\frac{\partial}{\partial y}$ *for* $n = 10$.

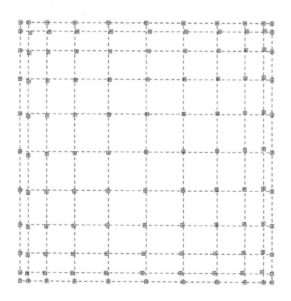

FIG. 2.2.5 - *The upwind grid corresponding to* $L = -\Delta + 2\frac{\partial^2}{\partial x \partial y}$ *for* $n = 10$.

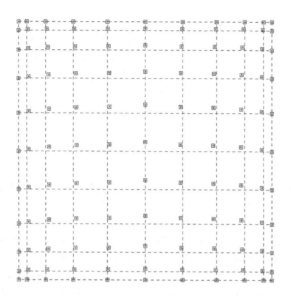

FIG. 2.2.6 - *The upwind grid corresponding to* $L = -\frac{1}{20}\Delta + x\frac{\partial}{\partial x} - \frac{\partial}{\partial y}$ *for* $n = 10$.

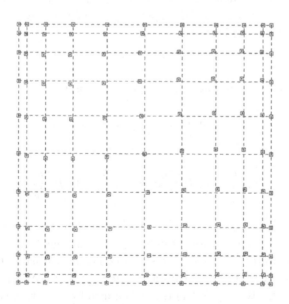

FIG. 2.2.7 - *The upwind grid corresponding to* $L = -\frac{1}{20}\Delta + y\frac{\partial}{\partial x} - x\frac{\partial}{\partial y}$ *for* $n = 10$.

For any fixed i and j, we compute by virtue of $(A.3.3)$, the quantities

$$(2.2.13) \qquad A_{ij} := r_n(\tau_{i,j}^{(n)}, \eta_i^{(n)}) = \sum_{k=0}^{n} r_n(\eta_k^{(n)}, \eta_i^{(n)}) \, \tilde{l}_k^{(n)}(\tau_{i,j}^{(n)}),$$

$$(2.2.14) \qquad B_{ij} := r_n(\eta_j^{(n)}, v_{i,j}^{(n)}) = \sum_{k=0}^{n} r_n(\eta_j^{(n)}, \eta_k^{(n)}) \, \tilde{l}_k^{(n)}(v_{i,j}^{(n)}).$$

Then, we evaluate

$$(2.2.15) \qquad C_{ij} := A_{ij} + B_{ij} - r_n(\eta_j^{(n)}, \eta_i^{(n)}) \approx r_n(\tau_{i,j}^{(n)}, v_{i,j}^{(n)}).$$

The determination of C_{ij} does not need all the $(n+1)^2$ values of r_n at the Legendre nodes, as the exact interpolation would require. Therefore, we will save computer time by using C_{ij} in place of $r_n(\tau_{i,j}^{(n)}, v_{i,j}^{(n)})$.

We denote by $\mathcal{S}_n$ the $(n+1)^2 \times (n+1)^2$ matrix that allows us to pass from the vector of the values of a polynomial r_n at the Legendre grid (the entries of the vector are ordered as suggested by $(1.2.3)$) to the corresponding values obtained by $(2.2.15)$, although these are not exactly the values at the upwind grid. The cost of implementing the multiplication of $\mathcal{S}_n$ by a vector is now proportional to n^3. Of course, if the upwind grid coincides with the Legendre grid, the matrix $\mathcal{S}_n$ is the identity.

2.3   The collocation method at the upwind grid

We approximate the solution of $(2.1.1)$-$(1.1.2)$ by seeking a polynomial $q_n \in \mathbf{P}_n^*$ such that

$$(2.3.1) \quad f_1 \frac{\partial^2 q_n}{\partial x^2} + f_2 \frac{\partial^2 q_n}{\partial x \partial y} + f_3 \frac{\partial^2 q_n}{\partial y^2} + f_4 \frac{\partial q_n}{\partial x} + f_5 \frac{\partial q_n}{\partial y} + f_6 q_n = f_7,$$

at the nodes $(\tau_{i,j}^{(n)}, v_{i,j}^{(n)})$, $1 \le i \le n-1$, $1 \le j \le n-1$, introduced in the previous section, and we impose the boundary conditions $(1.2.2)$. In the case of the Laplace operator $L = -\Delta$, the set of equations associated with $(2.3.1)$ coincides with those of the collocation problem $(1.2.1)$. When the operator L contains transport terms, $(2.3.1)$ is not the usual collocation scheme at the Legendre nodes, like in $(2.1.5)$. However, if $(2.2.12)$ holds true, we expect the convergence of q_n to the exact solution U. Although a theory is still not available, the convergence rate should be of spectral type.

There will be advantages to using the new collocation scheme. Actually, the approximated solutions obtained with the new grid look more accurate than those resulting from collocation at the Legendre grid, especially when the diffusion terms of L are very small compared to the transport terms. The upwind grid provides a better control of the highest Fourier coefficients of the expansion of the approximated solution in the Legendre basis functions. The relation (2.1.10) (or an approximated version of it) corresponds to requiring that the collocation equations are satisfied in the enlarged space $\mathbf{P}_n^* \oplus \{\lambda\chi_n, \ \lambda \in \mathbf{R}\}$, resulting in a relevant dumping effect on the high modes. Somehow, this corresponds to the introduction of a nonlinear *artificial viscosity* in the collocation scheme. A similar smoothing behavior was also pointed out in CANUTO (1994), and CANUTO and PUPPO (1994), where a different stabilization procedure, based on the *bubble function* approach, was introduced and analyzed. We remark that we do not have any theoretical background to justify the good influence of the upwind grid in the treatment of transport-dominated equations, although the tendency is confirmed by numerous experiments. Heuristic explanations and comparative results are given in FUNARO (1993) and FUNARO (1997). Other comparisons between approximated solutions obtained after collocation at the Legendre grid and the upwind grid, respectively, are reported in sections 2.4 and 6.1.

Another reason for preferring the new set of collocation nodes concerns the implementation of the method. As we saw in section 1.4, the only effective way to solve the linear systems arising from spectral collocation discretizations is to apply a preconditioned iterative procedure. In the case of the Laplace operator we were able to come out with a very good preconditioning matrix. In the case of transport-diffusion equations, the construction of an appropriate preconditioner is not trivial, and has been the subject of active research (see DEVILLE and MUND (1991), DEVILLE and MUND (1992), PINELLI, BENOCCI and DEVILLE (1994), and PINELLI, COUZY, DEVILLE and BENOCCI (1996)). For the collocation method at the Legendre grid, the use of naive finite-differences preconditioners does not yield, for transport-dominated equations, results comparable with the ones corresponding to the Poisson equation. The situation considerably improves for the collocation method at the upwind grid, where the finite-differences preconditioner can still be reconsidered with success. We are going to describe how the new algorithm will be implemented.

Once the nodes $(\tau_{i,j}^{(n)}, v_{i,j}^{(n)})$, $0 \le i \le n$, $0 \le j \le n$, have been determined, the set of equations (2.3.1)-(1.2.2) is equivalent to a linear system, and the corresponding $(n+1)^2 \times (n+1)^2$ matrix, because of the unstructured distribution of the upwind nodes in $\bar{\Omega}$, is completely full. It is clear that we are not interested in working directly with such a matrix. We first turn our attention to the spectral discretization of the operator L at the nodes

$(\eta_j^{(n)}, \eta_i^{(n)})$, $0 \le i \le n$, $0 \le j \le n$. The corresponding matrix, denoted by $\mathcal{L}_n$, transforms the vector $\vec{X}_n$ of the values of a polynomial $q_n \in \mathbf{P}_n^\star$ at the Legendre nodes, into the vector of the values of the function Lq_n at the same nodes. For $f_2 = 0$ the structure of $\mathcal{L}_n$ is the same as the one shown in Fig. 1.2.2 for $n = 4$. Arguing as in (1.2.4), the multiplication $\mathcal{L}_n \vec{X}_n$ can be carried out with a cost proportional to n^3. This is also true for $f_2 \ne 0$ (in this case $\mathcal{L}_n$ is completely full), if we recall that for any fixed $0 \le j \le n$ one has

$$(2.3.2) \qquad B_{ij} := \frac{\partial^2 q_n}{\partial x \partial y}(\eta_j^{(n)}, \eta_i^{(n)}) = \sum_{m=0}^n \vec{d}_{im}^{(1)} A_{mj} \qquad 0 \le i \le n,$$

$$\text{where} \qquad A_{mj} := \sum_{k=0}^n \vec{d}_{jk}^{(1)} q_n(\eta_k^{(n)}, \eta_m^{(n)}) \qquad 0 \le m \le n.$$

Therefore, we start by computing all the coefficients A_{mj}, $0 \le m \le n$ (with a cost proportional to n^2), then we evaluate B_{ij}, $0 \le i \le n$ (always with a cost proportional to n^2). Finally, we have to repeat this procedure $n + 1$ times as j goes from 0 up to n.

The next step is to pass from $(Lq_n)(\eta_j^{(n)}, \eta_i^{(n)})$ to $(Lq_n)(\tau_{i,j}^{(n)}, v_{i,j}^{(n)})$ for $0 \le i \le n$, $0 \le j \le n$. For this purpose, we will use the relation (2.2.15) that leads to the matrix $\mathcal{S}_n$. We denote by $\vec{F}_n$ the vector of dimension $(n + 1)^2$ containing the values of f_7 at the nodes $\Theta_k^{(n)}$, $0 \le k \le n_T$ (see (1.2.3)), and by $\vec{G}_n$ the vector introduced in section 1.7. We are able to treat properly the boundary conditions with the help of the notation of section 1.7. Hence, the collocation problem at the upwind grid leads to the following linear system:

$$(2.3.3) \qquad (\mathcal{J}_n^\Omega \mathcal{S}_n \mathcal{L}_n + \mathcal{J}_n^{\partial \Omega}) \vec{X}_n = \mathcal{J}_n^\Omega \mathcal{S}_n \vec{F}_n + \vec{G}_n.$$

Remember that the matrix $\mathcal{S}_n$ is not associated with the exact polynomial interpolation operator in $\mathbf{P}_n^\star$, as already pointed out in section 2.2. The modification was made in order to speed up the multiplication of $\mathcal{S}_n$ by a vector. Therefore, (2.3.3) does not exactly correspond to the collocation problem (2.3.1)-(1.1.2). Such a choice seems satisfactory, however, in view of the applications.

For the numerical solution of (2.3.3) we use the preconditioned Du Fort-Frankel method given in (1.4.6), where now $A_n := \mathcal{J}_n^\Omega \mathcal{S}_n \mathcal{L}_n + \mathcal{J}_n^{\partial \Omega}$ and $\vec{B}_n := \mathcal{J}_n^\Omega \mathcal{S}_n \vec{F}_n + \vec{G}_n$. We still have to explain how to build up the preconditioning matrix $\mathcal{B}_n$. Following section 1.4, we define $\mathcal{B}_n$ to be the matrix corresponding to the second-order finite-differences discretization of the differential operator L, centered at the upwind grid nodes. For any $1 \le i \le n - 1$, $1 \le j \le n - 1$, let $Q_{i,j} \in \mathbf{P}_2^\star$ be the second degree polynomial attaining the values $z_{i+\alpha,j+\beta}$ at the nine points $(\eta_{j+\beta}^{(n)}, \eta_{i+\alpha}^{(n)})$ respectively, where α and β belong to the set $\{-1,0,1\}$. Basically, the nine-points stencil considered is the one given by A_k, $1 \le k \le 9$, in Fig. 2.2.1.

$$
\begin{bmatrix}
1 & 0 & 0 & 0 & 0 & 0 & 0 & 0 & 0 & 0 & 0 & 0 & 0 & 0 & 0 & 0 & 0 & 0 & 0 & 0 & 0 & 0 & 0 & 0 & 0 & 0 & 0 \\
0 & 1 & 0 & 0 & 0 & 0 & 0 & 0 & 0 & 0 & 0 & 0 & 0 & 0 & 0 & 0 & 0 & 0 & 0 & 0 & 0 & 0 & 0 & 0 & 0 & 0 & 0 \\
0 & 0 & 1 & 0 & 0 & 0 & 0 & 0 & 0 & 0 & 0 & 0 & 0 & 0 & 0 & 0 & 0 & 0 & 0 & 0 & 0 & 0 & 0 & 0 & 0 & 0 & 0 \\
0 & 0 & 0 & 1 & 0 & 0 & 0 & 0 & 0 & 0 & 0 & 0 & 0 & 0 & 0 & 0 & 0 & 0 & 0 & 0 & 0 & 0 & 0 & 0 & 0 & 0 & 0 \\
0 & 0 & 0 & 0 & 1 & 0 & 0 & 0 & 0 & 0 & 0 & 0 & 0 & 0 & 0 & 0 & 0 & 0 & 0 & 0 & 0 & 0 & 0 & 0 & 0 & 0 & 0 \\
0 & 0 & 0 & 0 & 0 & 1 & 0 & 0 & 0 & 0 & 0 & 0 & 0 & 0 & 0 & 0 & 0 & 0 & 0 & 0 & 0 & 0 & 0 & 0 & 0 & 0 & 0 \\
\bullet & \bullet & \bullet & 0 & 0 & \bullet & \bullet & \bullet & 0 & 0 & \bullet & \bullet & \bullet & 0 & 0 & 0 & 0 & 0 & 0 & 0 & 0 & 0 & 0 & 0 & 0 & 0 & 0 \\
0 & \bullet & \bullet & \bullet & 0 & 0 & \bullet & \bullet & \bullet & 0 & 0 & \bullet & \bullet & \bullet & 0 & 0 & 0 & 0 & 0 & 0 & 0 & 0 & 0 & 0 & 0 & 0 & 0 \\
0 & 0 & \bullet & \bullet & \bullet & 0 & 0 & \bullet & \bullet & \bullet & 0 & 0 & \bullet & \bullet & \bullet & 0 & 0 & 0 & 0 & 0 & 0 & 0 & 0 & 0 & 0 & 0 & 0 \\
0 & 0 & 0 & 0 & 0 & 0 & 0 & 0 & 0 & 0 & 1 & 0 & 0 & 0 & 0 & 0 & 0 & 0 & 0 & 0 & 0 & 0 & 0 & 0 & 0 & 0 & 0 \\
0 & 0 & 0 & 0 & 0 & 0 & 0 & 0 & 0 & 0 & 0 & 1 & 0 & 0 & 0 & 0 & 0 & 0 & 0 & 0 & 0 & 0 & 0 & 0 & 0 & 0 & 0 \\
0 & 0 & 0 & 0 & 0 & \bullet & \bullet & \bullet & 0 & 0 & \bullet & \bullet & \bullet & 0 & 0 & \bullet & \bullet & \bullet & 0 & 0 & 0 & 0 & 0 & 0 & 0 & 0 & 0 \\
0 & 0 & 0 & 0 & 0 & 0 & \bullet & \bullet & \bullet & 0 & 0 & \bullet & \bullet & \bullet & 0 & 0 & \bullet & \bullet & \bullet & 0 & 0 & 0 & 0 & 0 & 0 & 0 & 0 \\
0 & 0 & 0 & 0 & 0 & 0 & 0 & \bullet & \bullet & \bullet & 0 & 0 & \bullet & \bullet & \bullet & 0 & 0 & \bullet & \bullet & \bullet & 0 & 0 & 0 & 0 & 0 & 0 & 0 \\
0 & 0 & 0 & 0 & 0 & 0 & 0 & 0 & 0 & 0 & 0 & 0 & 0 & 0 & 0 & 1 & 0 & 0 & 0 & 0 & 0 & 0 & 0 & 0 & 0 & 0 & 0 \\
0 & 0 & 0 & 0 & 0 & 0 & 0 & 0 & 0 & 0 & 0 & 0 & 0 & 0 & 0 & 0 & 1 & 0 & 0 & 0 & 0 & 0 & 0 & 0 & 0 & 0 & 0 \\
0 & 0 & 0 & 0 & 0 & 0 & 0 & 0 & 0 & 0 & \bullet & \bullet & \bullet & 0 & 0 & \bullet & \bullet & \bullet & 0 & 0 & \bullet & \bullet & \bullet & 0 & 0 & 0 & 0 \\
0 & 0 & 0 & 0 & 0 & 0 & 0 & 0 & 0 & 0 & 0 & \bullet & \bullet & \bullet & 0 & 0 & \bullet & \bullet & \bullet & 0 & 0 & \bullet & \bullet & \bullet & 0 & 0 & 0 \\
0 & 0 & 0 & 0 & 0 & 0 & 0 & 0 & 0 & 0 & 0 & 0 & \bullet & \bullet & \bullet & 0 & 0 & \bullet & \bullet & \bullet & 0 & 0 & \bullet & \bullet & \bullet & 0 & \bullet \\
0 & 0 & 0 & 0 & 0 & 0 & 0 & 0 & 0 & 0 & 0 & 0 & 0 & 0 & 0 & 0 & 0 & 0 & 0 & 0 & 1 & 0 & 0 & 0 & 0 & 0 & 0 \\
0 & 0 & 0 & 0 & 0 & 0 & 0 & 0 & 0 & 0 & 0 & 0 & 0 & 0 & 0 & 0 & 0 & 0 & 0 & 0 & 0 & 1 & 0 & 0 & 0 & 0 & 0 \\
0 & 0 & 0 & 0 & 0 & 0 & 0 & 0 & 0 & 0 & 0 & 0 & 0 & 0 & 0 & 0 & 0 & 0 & 0 & 0 & 0 & 0 & 1 & 0 & 0 & 0 & 0 \\
0 & 0 & 0 & 0 & 0 & 0 & 0 & 0 & 0 & 0 & 0 & 0 & 0 & 0 & 0 & 0 & 0 & 0 & 0 & 0 & 0 & 0 & 0 & 1 & 0 & 0 & 0 \\
0 & 0 & 0 & 0 & 0 & 0 & 0 & 0 & 0 & 0 & 0 & 0 & 0 & 0 & 0 & 0 & 0 & 0 & 0 & 0 & 0 & 0 & 0 & 0 & 1 & 0 & 0 \\
0 & 0 & 0 & 0 & 0 & 0 & 0 & 0 & 0 & 0 & 0 & 0 & 0 & 0 & 0 & 0 & 0 & 0 & 0 & 0 & 0 & 0 & 0 & 0 & 0 & 1 & 0 \\
0 & 0 & 0 & 0 & 0 & 0 & 0 & 0 & 0 & 0 & 0 & 0 & 0 & 0 & 0 & 0 & 0 & 0 & 0 & 0 & 0 & 0 & 0 & 0 & 0 & 0 & 1
\end{bmatrix}
$$

FIG. 2.3.1 - *Structure of the preconditioning matrix B_4.*

Successively, we generalize (1.4.1) and (1.4.3) by combining the set of values $\{z_{i,j}\}_{\substack{0 \le i \le n \\ 0 \le j \le n}}$ in order to satisfy the following equations:

$$(2.3.4) \qquad f_1(\eta_j^{(n)}, \eta_i^{(n)})\frac{\partial^2 Q_{i,j}}{\partial x^2}(\tau_{i,j}^{(n)}, \upsilon_{i,j}^{(n)}) + f_2(\eta_j^{(n)}, \eta_i^{(n)})\frac{\partial^2 Q_{i,j}}{\partial x \partial y}(\tau_{i,j}^{(n)}, \upsilon_{i,j}^{(n)})$$

$$+ f_3(\eta_j^{(n)}, \eta_i^{(n)})\frac{\partial^2 Q_{i,j}}{\partial y^2}(\tau_{i,j}^{(n)}, \upsilon_{i,j}^{(n)}) + f_4(\eta_j^{(n)}, \eta_i^{(n)})\frac{\partial Q_{i,j}}{\partial x}(\tau_{i,j}^{(n)}, \upsilon_{i,j}^{(n)})$$

$$+ f_5(\eta_j^{(n)}, \eta_i^{(n)})\frac{\partial Q_{i,j}}{\partial y}(\tau_{i,j}^{(n)}, \upsilon_{i,j}^{(n)}) + f_6(\eta_j^{(n)}, \eta_i^{(n)})Q_{i,j}(\tau_{i,j}^{(n)}, \upsilon_{i,j}^{(n)})$$

$$= f_7(\tau_{i,j}^{(n)}, \upsilon_{i,j}^{(n)}), \quad 1 \le i \le n-1, \ 1 \le j \le n-1,$$

together with the boundary conditions in (1.4.2). A similar version is obtained by evaluating the functions f_k, $1 \le k \le 6$, in (2.3.4) at the upwind nodes. We prefer the first version, since the coefficients of the differential operator are usually known at the Legendre nodes (the reader may think to the case of a nonlinear equation where the coefficients are a function of the unknown, as in section 5.1). The relations above allow the construction of $\mathcal{B}_n$. The explicit expression of the entries is given in FUNARO (1993). We discover that $\mathcal{B}_n$ is banded with bandwidth equal to $2n + 5$. The structure of $\mathcal{B}_4$ can be found in Fig. 2.3.1. Finite-differences approximations of transport-diffusion equations based on the upwind grid (thus, using $\mathcal{B}_n$ as discretization matrix in place of $\mathcal{A}_n$) have been successfully experimented in FUNARO and RUSSO (1993). When $L = -\Delta$, then the upwind grid coincides with the Legendre grid and the matrix $\mathcal{B}_n$ turns out to be the same already considered in section 1.4.

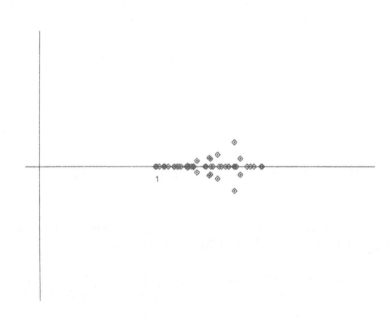

FIG. 2.3.2 - *Eigenvalues of* $\mathcal{B}_8^{-1}\mathcal{A}_8$ *for* $\epsilon = .1$ *and* $\vec{\beta} \equiv (1,0)$ *in* (2.1.4).

We recall that, using the fact that $\mathcal{B}_n$ is banded, we are able to find its *LU* decomposition with a cost proportional to n^4. In addition, we would like to have it that $\mathcal{B}_n^{-1}\mathcal{A}_n \approx \mathcal{I}_n$, where $\mathcal{I}_n$ is the identity matrix, which means that $\mathcal{B}_n$ is a good preconditioner. Actually, for constant coefficients differential operators, $\mathcal{A}_n$ and $\mathcal{B}_n$ behave in the same way on the polynomials of the subspace $\mathbf{P}_2^\star \subset \mathbf{P}_n^\star$, which implies that 1 is eigenvalue of $\mathcal{B}_n^{-1}\mathcal{A}_n$. The matrices $\mathcal{A}_n$ and $\mathcal{B}_n$ show a similar behavior also for the high modes, so that the eigenvalues of $\mathcal{B}_n^{-1}\mathcal{A}_n$ are expected to be very well distributed.

We studied the spectrum of $\mathcal{B}_n^{-1}\mathcal{A}_n$ in the case of the equation (2.1.4). Typical configurations of the preconditioned eigenvalues in the complex plane **C** are given in Figs. 2.3.2 and 2.3.3, where for $n = 8$ and $\epsilon = .1$ we considered $\vec{\beta} = (1,0)$ and $\vec{\beta} = (1,2)$, respectively. The eigenvalues are complex, presenting however a small imaginary part. They are clustered around the interval $[1,2] \subset$ **C**. We guess that the distribution would be even better if the matrix $\mathcal{S}_n$ were the exact interpolation operator in the space $\mathbf{P}_n^\star$. Actually, the situation improves when the flux $\epsilon^{-1}\vec{\beta}$ has the same orientation as one of the coordinate axes x or y (note that in these special cases we have $C_{ij} = r_n(\tau_{i,j}^{(n)}, \upsilon_{i,j}^{(n)})$ in (2.2.15)). The behavior does not substantially change by increasing n or diminishing ϵ.

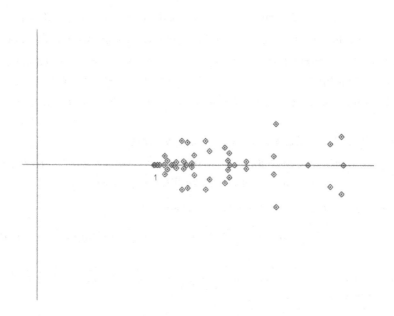

FIG. 2.3.3 - *Eigenvalues of $\mathcal{B}_8^{-1}\mathcal{A}_8$ for $\epsilon = .1$ and $\vec{\beta} \equiv (1,2)$ in (2.1.4).*

If we want to implement the Du Fort-Frankel method, we need some a priori estimate of the magnitude of the eigenvalues. We take σ_1 and σ_2 as in (1.4.9), where $\mu^{(n)}$ is given by (1.4.4), after defining $\phi_n \in \mathbf{P}_n^*$ such that $\phi_n(\eta_j^{(n)}, \eta_i^{(n)}) := (-1)^{i+j}[1 + f_4(\eta_j^{(n)}, \eta_i^{(n)})\eta_j^{(n)} + f_5(\eta_j^{(n)}, \eta_i^{(n)})\eta_i^{(n)}]$. This choice of parameters is not the optimal one. However, in all the experiments we considered, σ_1 and σ_2 were always in the stability region of the method (1.4.6) and not too far from the optimal values. In fact, the convergence up to the machine accuracy was achieved in less than 30 iterations. The procedure seems to be reliable also for equations with non-constant coefficients. Several experiments are provided in the next section.

2.4 The numerical algorithm for the transport-diffusion equation

As in section 1.5, we outline the basic steps to be followed for the determination of the approximated solution to problem (2.1.1)-(1.1.2):

1) Compute the nodes $\eta_i^{(n)}$, $0 \le i \le n$, the weights $\tilde{w}_i^{(n)}$, $0 \le i \le n$ and the entries of the differentiation matrices $\tilde{D}_n$ and $\tilde{D}_n^2$;

2) Compute the nodes $(\tau_{i,j}^{(n)}, \upsilon_{i,j}^{(n)})$, $0 \le i \le n$, $0 \le j \le n$, of the upwind grid, following the instructions of section 2.2;

3) Construct the banded matrix $\mathcal{B}_n$ (see (2.3.4) and FUNARO (1993));

4) Compute the LU factorization of $\mathcal{B}_n$, using algorithms for banded matrices;

5) Construct the right-hand side vector $\vec{B}_n \equiv \mathcal{J}_n^\Omega \mathcal{S}_n \vec{F}_n + \vec{G}_n$ in (2.3.3), using the shifting operator $\mathcal{S}_n$ introduced at the end of section 2.2;

6) Solve the system $\mathcal{B}_n \vec{Z}_n = \vec{B}_n$ with the help of the LU decomposition of $\mathcal{B}_n$;

7) Set $\vec{X}_n^0 \equiv \vec{X}_n^1 \equiv \vec{Z}_n$ to be the starting vectors;

8) Use the vectors $\vec{\Phi}_n$ and $\vec{\Psi}_n$ as described in sections 1.4 and 2.3 to determine $\mu^{(n)}$ in (1.4.4) and the values of σ_1 and σ_2 in (1.4.9).

Successively, for $k \ge 1$, we proceed as follows:

9) Compute $\mathcal{L}_n \vec{X}_n^k$ and $\mathcal{S}_n(\mathcal{L}_n \vec{X}_n^k - \vec{F}_n)$, using the differentiation matrices $\tilde{D}_n$ and $\tilde{D}_n^2$;

10) Impose the boundary conditions by writing $\mathcal{J}_n^\Omega \mathcal{S}_n(\mathcal{L}_n \vec{X}_n^k - \vec{F}_n) = \mathcal{A}_n \vec{X}_n^k - \vec{B}_n$, where $\mathcal{A}_n = \mathcal{J}_n^\Omega \mathcal{S}_n \mathcal{L}_n + \mathcal{J}_n^{\partial\Omega}$;

11) Compute $\mathcal{B}_n^{-1}(\mathcal{A}_n \vec{X}_n^k - \vec{B}_n)$ using the LU decomposition of $\mathcal{B}_n$;

12) Update $\vec{X}_n^{k+1}$ according to (1.4.6), with σ_1 and σ_2 chosen as in step 8;

13) Check the residual $\|A_n \vec{X}_n^{k+1} - \vec{B}_n\|$, where $\|\cdot\|$ is a suitable norm (for instance the one in (1.4.5)). If this is less than a prescribed error then stop, otherwise go back to step 9.

In the output vector, we find the values of the approximating polynomial at the nodes $(\eta_j^{(n)}, \eta_i^{(n)})$, $0 \leq i \leq n$, $0 \leq j \leq n$.

We now present the results of some numerical tests. In equation (2.1.4) we set $f_7 = 1$ and $\epsilon = \frac{1}{10}$. Homogeneous Dirichlet boundary conditions are assumed on $\partial\Omega$. The approximated solutions for $n = 10$ are shown in Figs. 2.4.1 and 2.4.2 for $\vec{\beta} = (1,0)$ and $\vec{\beta} = (1,\frac{1}{2})$, respectively. The grids corresponding to these experiments are similar to those given in Figs. 2.2.2 and 2.2.4. With respect to Fig. 1.5.1, the solutions are *transported* in the direction of $\vec{\beta}$, and boundary layers are developing at the outflow boundaries.

By diminishing the value of the parameter ϵ, the gradients get sharper. We give in Figs. 2.4.3 and 2.4.4 the approximating polynomials corresponding to $\epsilon = \frac{1}{50}$, with $n = 20$, for $\vec{\beta} = (1,0)$ and $\vec{\beta} = (1,\frac{1}{2})$, respectively.

In all these experiments the degree n was sufficient to correctly resolve the boundary layers. Let us now see what happens with a lower degree n when, for example, $\epsilon = \frac{1}{50}$. We compare for $n = 10$ the solutions obtained by the collocation method at the upwind grid (Figs. 2.4.5 and 2.4.6) with those obtained after collocation at the Legendre grid (Figs. 2.4.7 and 2.4.8). The difference between the two approaches is clear and it is even more evident when n is smaller. We will be able to add more comments in section 6.1. We also recall that the iterations needed to evaluate the solution of the collocation problem at the upwind grid are less than those necessary in the other case, since we can rely on a much better preconditioner.

In the case of the vector $\vec{\beta} = (1, \frac{1}{2})$, which is not parallel to any of the coordinate axes, the computation of the approximated solutions is a little bit more difficult. The number of iterations of the preconditioned Du Fort-Frankel method varies depending on the magnitude of n and ϵ, never exceeding however an acceptable threshold.

We present the last experiment of this chapter, which is the boundary-value problem:

$$(2.4.1) \quad \begin{cases} -\frac{1}{1000}\Delta U + (1+y)\dfrac{\partial U}{\partial x} - (1+x)\dfrac{\partial U}{\partial y} = 0 & \text{in } \Omega, \\[2mm] U(-1,y) = g_4(y) = 1 - y^2 & y \in]-1,1[, \\[2mm] U = 0 & \text{on the other three sides of } \partial\Omega. \end{cases}$$

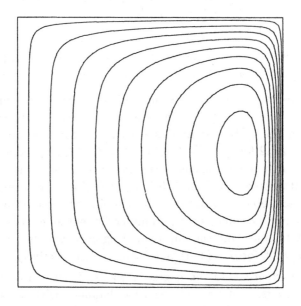

FIG. 2.4.1 - *Approximated solution for* $n = 10$, $\epsilon = \frac{1}{10}$ *and* $\vec{\beta} \equiv (1, 0)$.

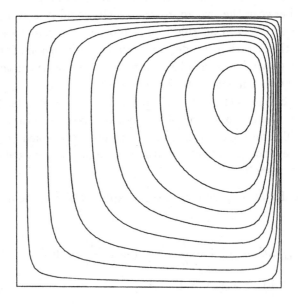

FIG. 2.4.2 - *Approximated solution for* $n = 10$, $\epsilon = \frac{1}{10}$ *and* $\vec{\beta} \equiv (1, \frac{1}{2})$.

FIG. 2.4.3 - *Approximated solution for* $n = 20$, $\epsilon = \frac{1}{50}$ *and* $\vec{\beta} \equiv (1, 0)$.

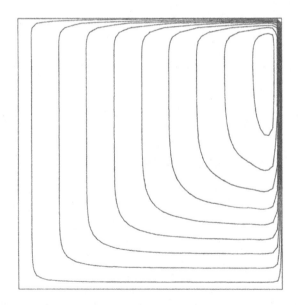

FIG. 2.4.4 - *Approximated solution for* $n = 20$, $\epsilon = \frac{1}{50}$ *and* $\vec{\beta} \equiv (1, \frac{1}{2})$.

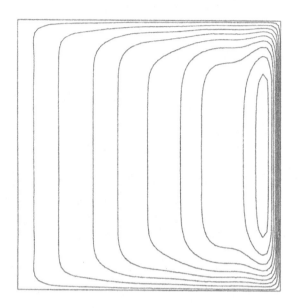

FIG. 2.4.5 - *Approximated solution for $n = 10$, $\epsilon = \frac{1}{50}$ and $\vec{\beta} \equiv (1, 0)$ (collocation at the upwind grid – solution underresolved but stabilized).*

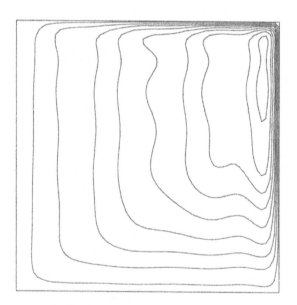

FIG. 2.4.6 - *Approximated solution for $n = 10$, $\epsilon = \frac{1}{50}$ and $\vec{\beta} \equiv (1, \frac{1}{2})$ (collocation at the upwind grid – solution underresolved but stabilized).*

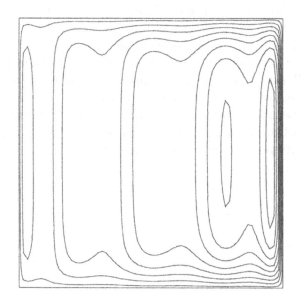

FIG. 2.4.7 - *Approximated solution for $n = 10$, $\epsilon = \frac{1}{50}$ and $\vec{\beta} \equiv (1,0)$ (collocation at the Legendre grid – solution presenting oscillations).*

FIG. 2.4.8 - *Approximated solution for $n = 10$, $\epsilon = \frac{1}{50}$ and $\vec{\beta} \equiv (1,\frac{1}{2})$ (collocation at the Legendre grid – solution presenting oscillations).*

The boundary condition assigned on the side $\{-1\} \times \,] - 1, 1[$ is "transported" on the side $\,] - 1, 1[\times \{-1\}$ along the circumferences centered at $(-1, -1)$. A similar vector field $\vec{\beta}$ will be considered in section 5.1.

The contour lines of the approximating polynomial of degree $n = 30$, satisfying the collocation problem at the upwind grid, are shown in Fig. 2.4.9. Some wiggles tell us that an accurate resolution of the boundary layers is still not achieved. We also remark that the exact solution U presents a sharp variation of the gradient along the curve $(x+1)^2 + (y+1)^2 = 4$, $(x, y) \in \Omega$, which negatively affects the accuracy of the computation.

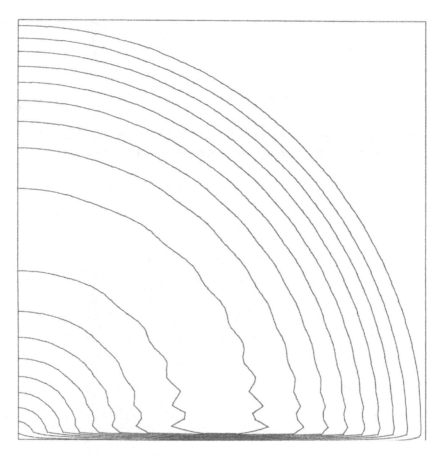

FIG. 2.4.9 - *Approximated solution to problem* (2.4.1) *for* $n = 30$.

3

Other Kinds of
Boundary Conditions

We study the polynomial approximation of boundary-value problems of elliptic
type subjected to Neumann or mixed-type boundary conditions. As we did in
the previous chapters, we develop a preconditioned iterative method for a fast
and accurate resolution of the corresponding linear systems.

3.1 Neumann-type boundary conditions

We look for a solution $U : \bar{\Omega} \rightarrow \mathbf{R}$ to the equation (2.1.1), but this time
we replace the Dirichlet boundary conditions (1.1.2) with *Neumann* boundary
conditions, i.e. we assign the values of $\frac{\partial}{\partial \bar{\nu}} U$ on $\partial \Omega$, where $\bar{\nu}$ is the outward
normal vector to Ω. In practice, we have

(3.1.1)
$$
\begin{cases}
-\dfrac{\partial U}{\partial y}(x,-1) = h_1(x) & \forall x \in]-1,1[, \\[2em]
\dfrac{\partial U}{\partial x}(1,y) = h_2(y) & \forall y \in]-1,1[, \\[2em]
\dfrac{\partial U}{\partial y}(x,1) = h_3(x) & \forall x \in]-1,1[, \\[2em]
-\dfrac{\partial U}{\partial x}(-1,y) = h_4(y) & \forall y \in]-1,1[,
\end{cases}
$$

and at the corners we set

$$(3.1.2) \quad \begin{cases} \left(-\dfrac{\partial U}{\partial x} - \dfrac{\partial U}{\partial y}\right)(-1,-1) = h_1(-1) + h_4(-1), \\[2mm] \left(\dfrac{\partial U}{\partial x} - \dfrac{\partial U}{\partial y}\right)(1,-1) = h_1(1) + h_2(-1), \\[2mm] \left(\dfrac{\partial U}{\partial x} + \dfrac{\partial U}{\partial y}\right)(1,1) = h_2(1) + h_3(1), \\[2mm] \left(-\dfrac{\partial U}{\partial x} + \dfrac{\partial U}{\partial y}\right)(-1,1) = h_3(-1) + h_4(1). \end{cases}$$

The functions $h_k : [-1,1] \to \mathbf{R}$, $1 \le k \le 4$ are given, and they are supposed to be continuous.

The case in which $f_1 = f_3 = -1$, $f_2 = f_4 = f_5 = 0$ and $f_6 = \lambda \ge 0$ in (2.1.1) corresponds to the *Helmholtz*-type equation

$$(3.1.3) \qquad\qquad -\Delta U + \lambda U = f \quad \text{in } \Omega.$$

Concerning this particular equation, existence and uniqueness are guaranteed by the following theorem.

THEOREM 3.1.1 - *Let f be a continuous function in Ω satisfying $\int_\Omega f^2\,dxdy$ $< +\infty$, and let h_k, $1 \le k \le 4$, be continuous functions in $[-1,1]$ with $h_1(1) = h_2(-1)$, $h_2(1) = h_3(1)$, $h_3(-1) = h_4(1)$, $h_4(-1) = h_1(-1)$. Then, if $\lambda > 0$ there exists a unique solution $U \in C^0(\bar\Omega) \cap C^1(\Omega)$ of (3.1.1)-(3.1.2)-(3.1.3). If $\lambda = 0$, there exist solutions only if the compatibility condition*

$$(3.1.4) \qquad \int_\Omega f\,dxdy + \int_{-1}^1 (h_1 + h_3)dx + \int_{-1}^1 (h_4 + h_2)dy = 0$$

is fullfilled. In this case there is no uniqueness, but the difference between two solutions is constant.

The solution U to (3.1.1)-(3.1.2)-(3.1.3) also satisfies

$$(3.1.5) \qquad B(U,\phi) := \int_\Omega \left(\frac{\partial U}{\partial x}\frac{\partial \phi}{\partial x} + \frac{\partial U}{\partial y}\frac{\partial \phi}{\partial y}\right) dxdy + \lambda \int_\Omega U\phi\, dxdy$$

$$= \int_\Omega (-\Delta U + \lambda U)\phi\, dxdy - \int_{-1}^1 \left(\frac{\partial U}{\partial y}\phi\right)(x,-1)\, dx + \int_{-1}^1 \left(\frac{\partial U}{\partial x}\phi\right)(1,y)\, dy$$

$$+ \int_{-1}^{1} \left(\frac{\partial U}{\partial y} \phi \right) (x, 1) \, dx \;-\; \int_{-1}^{1} \left(\frac{\partial U}{\partial x} \phi \right) (-1, y) \, dy$$

$$= \int_{\Omega} f\phi \, dx dy \;+\; \int_{-1}^{1} h_1(x)\phi(x, -1) \, dx \;+\; \int_{-1}^{1} h_2(y)\phi(1, y) \, dy$$

$$+ \int_{-1}^{1} h_3(x)\phi(x, 1) \, dx \;+\; \int_{-1}^{1} h_4(y)\phi(-1, y) \, dy \;=:\; F(\phi),$$

for any $\phi \in C^0(\bar{\Omega}) \cap C^1(\Omega)$.

To obtain (3.1.5), we multiplied (3.1.3) by a test function ϕ, integrated in Ω and we used the Green formulas. One may check that a regular function U satisfying (3.1.5) is also solution to (3.1.1)-(3.1.2)-(3.1.3). In addition, when $h_k = 0$, $1 \le k \le 4$, we can show the estimate (1.1.7).

For a theoretical analysis of elliptic equations with Neumann boundary conditions we refer to BITSADZE (1968) or DENNEMEYER (1968). An analysis of existence and uniqueness in a weak sense (i.e. starting from the variational formulation (3.1.5)) can be carried out using theorem 1.1.2, embedding the solution in a suitable functional space **X** (see BREZZI and GILARDI (1987)).

We soon examine the numerical approximation. For $n \ge 2$, the collocation method applied to our set of equations (3.1.1)-(3.1.2)-(3.1.3) amounts to finding a polynomial $q_n \in \mathbf{P}_n^*$ such that

$$(3.1.6) \qquad (-\Delta q_n + \lambda q_n)(\eta_j^{(n)}, \eta_i^{(n)}) \;=\; f(\eta_j^{(n)}, \eta_i^{(n)}),$$

for $1 \le i \le n - 1$, $1 \le j \le n - 1$.

For what concerns the boundary conditions, we impose

$$(3.1.7) \quad
\begin{cases}
-\dfrac{\partial q_n}{\partial y}(\eta_j^{(n)}, \eta_0^{(n)}) = h_1(\eta_j^{(n)}) & 1 \le j \le n - 1, \\[2ex]
\dfrac{\partial q_n}{\partial x}(\eta_n^{(n)}, \eta_i^{(n)}) = h_2(\eta_i^{(n)}) & 1 \le i \le n - 1, \\[2ex]
\dfrac{\partial q_n}{\partial y}(\eta_j^{(n)}, \eta_n^{(n)}) = h_3(\eta_j^{(n)}) & 1 \le j \le n - 1, \\[2ex]
-\dfrac{\partial q_n}{\partial x}(\eta_0^{(n)}, \eta_i^{(n)}) = h_4(\eta_i^{(n)}) & 1 \le i \le n - 1,
\end{cases}$$

$$(3.1.8) \quad \begin{cases} \left(-\dfrac{\partial q_n}{\partial x} - \dfrac{\partial q_n}{\partial y}\right)(\eta_0^{(n)}, \eta_0^{(n)}) = h_1(\eta_0^{(n)}) + h_4(\eta_0^{(n)}), \\[2mm] \left(\dfrac{\partial q_n}{\partial x} - \dfrac{\partial q_n}{\partial y}\right)(\eta_n^{(n)}, \eta_0^{(n)}) = h_1(\eta_n^{(n)}) + h_2(\eta_0^{(n)}), \\[2mm] \left(\dfrac{\partial q_n}{\partial x} + \dfrac{\partial q_n}{\partial y}\right)(\eta_n^{(n)}, \eta_n^{(n)}) = h_2(\eta_n^{(n)}) + h_3(\eta_n^{(n)}), \\[2mm] \left(-\dfrac{\partial q_n}{\partial x} + \dfrac{\partial q_n}{\partial y}\right)(\eta_0^{(n)}, \eta_n^{(n)}) = h_3(\eta_0^{(n)}) + h_4(\eta_n^{(n)}). \end{cases}$$

We expect that q_n converges to U in a spectral way. Actually, this is what one can observe in the numerical experiments. Nevertheless, an analysis of convergence has not yet been developed. We will see in the next section how to modify (3.1.7) and (3.1.8) in order to present an alternative way to discretize the Neumann boundary conditions. Together with the possibility of developing a theoretical analysis, this new approach will be preferable for several other reasons.

3.2 Boundary conditions in weak form

We go back for a moment to the results in chapter one. For the problem (1.1.1)-(1.1.2) with homogeneous Dirichlet boundary conditions (i.e., $g_k = 0$, $1 \leq k \leq 4$), we recall the variational formulation (1.1.6), which uses test functions $\phi \in C^0(\bar{\Omega}) \cap C^1(\Omega)$ vanishing on $\partial\Omega$. By replacing the integrals with the sums and using $\mathbf{P}_n^{\star,0}$ as space of test functions, we end up with (1.3.6) which characterizes the approximated solution $s_n \in \mathbf{P}_n^{\star,0}$. The analysis of convergence of s_n to U for $n \to +\infty$, developed in theorem 1.3.3, is strongly based on the fact that both s_n and U are solutions to problems written in variational form, namely (1.1.6) and (1.3.6), respectively.

We can repeat the same trick for the case of Neumann boundary conditions. Hence, we start from (3.1.5) and we replace the integrals with the sums. Thus, the approximated solution $q_n \in \mathbf{P}_n^\star$ is required to satisfy

$$(3.2.1) \quad \sum_{i,j=0}^{n} (\vec{\nabla} q_n \cdot \vec{\nabla}\phi)(\eta_j^{(n)}, \eta_i^{(n)})\, \tilde{w}_i^{(n)} \tilde{w}_j^{(n)} + \lambda \sum_{i,j=0}^{n} (q_n \phi)(\eta_j^{(n)}, \eta_i^{(n)})\, \tilde{w}_i^{(n)} \tilde{w}_j^{(n)}$$

$$= \sum_{i,j=0}^{n} [(-\Delta q_n + \lambda q_n)\phi](\eta_j^{(n)}, \eta_i^{(n)})\, \tilde{w}_i^{(n)} \tilde{w}_j^{(n)}$$

$$- \sum_{j=0}^{n} \left(\frac{\partial q_n}{\partial y} \phi \right) (\eta_j^{(n)}, \eta_0^{(n)}) \tilde{w}_j^{(n)} + \sum_{i=0}^{n} \left(\frac{\partial q_n}{\partial x} \phi \right) (\eta_n^{(n)}, \eta_i^{(n)}) \tilde{w}_i^{(n)}$$

$$+ \sum_{j=0}^{n} \left(\frac{\partial q_n}{\partial y} \phi \right) (\eta_j^{(n)}, \eta_n^{(n)}) \tilde{w}_j^{(n)} - \sum_{i=0}^{n} \left(\frac{\partial q_n}{\partial x} \phi \right) (\eta_0^{(n)}, \eta_i^{(n)}) \tilde{w}_i^{(n)}$$

$$= \sum_{i,j=0}^{n} (f\phi)(\eta_j^{(n)}, \eta_i^{(n)}) \, \tilde{w}_i^{(n)} \tilde{w}_j^{(n)}$$

$$+ \sum_{j=0}^{n} h_1(\eta_j^{(n)}) \phi(\eta_j^{(n)}, \eta_0^{(n)}) \tilde{w}_j^{(n)} + \sum_{i=0}^{n} h_2(\eta_i^{(n)}) \phi(\eta_n^{(n)}, \eta_i^{(n)}) \tilde{w}_i^{(n)}$$

$$+ \sum_{j=0}^{n} h_3(\eta_j^{(n)}) \phi(\eta_j^{(n)}, \eta_n^{(n)}) \tilde{w}_j^{(n)} + \sum_{i=0}^{n} h_4(\eta_i^{(n)}) \phi(\eta_0^{(n)}, \eta_i^{(n)}) \tilde{w}_i^{(n)},$$

for any $\phi \in \mathbf{P}_n^\star$.
In the above equations, we took into account the quadrature formula $(A.2.3)$ and we integrated by parts as we did in obtaining $(1.3.6)$. If we choose $\phi \in \mathbf{P}_n^{\star,0}$, we have

$$(3.2.2) \qquad \sum_{i,j=0}^{n} [(-\Delta q_n + \lambda q_n)\phi](\eta_j^{(n)}, \eta_i^{(n)}) \, \tilde{w}_i^{(n)} \tilde{w}_j^{(n)}$$

$$= \sum_{i,j=0}^{n} (f\phi)(\eta_j^{(n)}, \eta_i^{(n)}) \, \tilde{w}_i^{(n)} \tilde{w}_j^{(n)}.$$

This relation implies $(3.1.6)$ simply by taking $\phi(x, y) = \tilde{l}_i^{(n)}(x)\tilde{l}_j^{(n)}(y)$, $(x, y) \in \bar{\Omega}$, $1 \leq i \leq n - 1$, $1 \leq j \leq n - 1$, where $\tilde{l}_k^{(n)}$, $0 \leq k \leq n$, are the Lagrange polynomials with respect to the Legendre nodes (see $(A.3.1)$). Therefore, at the nodes inside Ω, we find out the usual collocation scheme for the equation $(3.1.3)$. Something different happens at the boundary. In fact, if $\phi(x, y) = \tilde{l}_j^{(n)}(x)\tilde{l}_0^{(n)}(y)$, $(x, y) \in \bar{\Omega}$, $1 \leq j \leq n - 1$, one gets

$$(3.2.3) \qquad (-\Delta q_n + \lambda q_n)(\eta_j^{(n)}, \eta_0^{(n)}) - \frac{1}{\tilde{w}_0^{(n)}} \frac{\partial q_n}{\partial y}(\eta_j^{(n)}, \eta_0^{(n)})$$

$$= f(\eta_j^{(n)}, \eta_0^{(n)}) + \frac{1}{\tilde{w}_0^{(n)}} h_1(\eta_j^{(n)}), \qquad 1 \leq j \leq n - 1,$$

where we divided by $\tilde{w}_j^{(n)} \tilde{w}_0^{(n)}$. Similarly, at the other sides of the boundary, a suitable choice of the test functions in $\mathbf{P}_n^\star$ yields

$$(3.2.4) \qquad (-\Delta q_n + \lambda q_n)(\eta_n^{(n)}, \eta_i^{(n)}) + \frac{1}{\tilde{w}_n^{(n)}} \frac{\partial q_n}{\partial x}(\eta_n^{(n)}, \eta_i^{(n)})$$

$$= f(\eta_n^{(n)}, \eta_i^{(n)}) + \frac{1}{\tilde{w}_n^{(n)}} h_2(\eta_i^{(n)}), \qquad 1 \le i \le n-1,$$

$$(3.2.5) \qquad (-\Delta q_n + \lambda q_n)(\eta_j^{(n)}, \eta_n^{(n)}) + \frac{1}{\tilde{w}_n^{(n)}} \frac{\partial q_n}{\partial y}(\eta_j^{(n)}, \eta_n^{(n)})$$

$$= f(\eta_j^{(n)}, \eta_n^{(n)}) + \frac{1}{\tilde{w}_n^{(n)}} h_3(\eta_j^{(n)}), \qquad 1 \le j \le n-1,$$

$$(3.2.6) \qquad (-\Delta q_n + \lambda q_n)(\eta_0^{(n)}, \eta_i^{(n)}) - \frac{1}{\tilde{w}_0^{(n)}} \frac{\partial q_n}{\partial x}(\eta_0^{(n)}, \eta_i^{(n)})$$

$$= f(\eta_0^{(n)}, \eta_i^{(n)}) + \frac{1}{\tilde{w}_0^{(n)}} h_4(\eta_i^{(n)}), \qquad 1 \le i \le n-1.$$

At the corners we find combinations of the expressions above. For example, taking $\phi(x,y) = \tilde{l}_0^{(n)}(x) \tilde{l}_0^{(n)}(y)$, $(x,y) \in \bar{\Omega}$, at the point $(-1,-1)$ one has

$$(3.2.7) \quad (-\Delta q_n + \lambda q_n)(\eta_0^{(n)}, \eta_0^{(n)}) + \frac{1}{\tilde{w}_0^{(n)}} \left(-\frac{\partial q_n}{\partial x} - \frac{\partial q_n}{\partial y} \right) (\eta_0^{(n)}, \eta_0^{(n)})$$

$$= f(\eta_0^{(n)}, \eta_0^{(n)}) + \frac{1}{\tilde{w}_0^{(n)}} h_1(\eta_0^{(n)}) + \frac{1}{\tilde{w}_0^{(n)}} h_4(\eta_0^{(n)}).$$

Similar relations hold at the nodes $(1,-1)$, $(-1,1)$, $(1,1)$. Finally, we note that by virtue of $(A.2.4)$, we have $[\tilde{w}_0^{(n)}]^{-1} = [\tilde{w}_n^{(n)}]^{-1} = \frac{1}{2}n(n+1)$.

Between the way of imposing the boundary conditions for q_n proposed in the section 3.1 and the one here considered there is an attempt to collocate the differential equation (3.1.3) also at the boundary nodes. At the same time, we may assume that the collocation is made at all the nodes in $\Omega \cup \partial\Omega$ and that the boundary conditions are interpreted as a sort of *penalty* terms. A drawback is that we also need to know f on $\partial\Omega$, which is not necessary when using (3.1.7)-(3.1.8).

We will say that the approximated solutions obtained with the new approach satisfy the Neumann boundary conditions in a *weak sense*. The first advantage of weakly imposing the Neumann conditions is that they easily lead to the variational formulation (3.2.1), so that proving the convergence of q_n

to U is now an easier exercise. For U sufficiently regular, when $n \to +\infty$, we expect that the residual $-\Delta q_n + \lambda q_n - f$ also tends to zero at the boundary, so that q_n tends to satisfy the Neumann conditions in a *strong sense* as in (3.1.7)-(3.1.8). Details about the convergence proof are given in BERNARDI and MADAY (1992).

Another positive aspect of the weak formulation is that it produces better numerical results than the strong formulation. Comparative tests are discussed in FUNARO (1992) p. 205, and QUARTERONI and ZAMPIERI (1992). We note that the cost of implementing the collocation scheme imposing the boundary conditions either in the weak or in the strong sense, is exactly the same. For instance, using the differentiation matrices $\tilde{D}_n$ and $\tilde{D}_n^2$ and recalling (1.2.4), we can express (3.2.3) as follows:

$$(3.2.8) \quad -\sum_{m=0}^{n} \left[\tilde{d}_{jm}^{(2)} \, q_n(\eta_m^{(n)}, \eta_0^{(n)}) + \tilde{d}_{0m}^{(2)} \, q_n(\eta_j^{(n)}, \eta_m^{(n)}) \right] + \lambda q_n(\eta_j^{(n)}, \eta_0^{(n)})$$

$$-\frac{1}{\tilde{w}_0^{(n)}} \sum_{m=0}^{n} \tilde{d}_{0m}^{(1)} \, q_n(\eta_j^{(n)}, \eta_m^{(n)}) = f(\eta_j^{(n)}, \eta_0^{(n)}) + \frac{1}{\tilde{w}_0^{(n)}} \, h_1(\eta_j^{(n)}),$$

for $1 \leq j \leq n - 1$. In a similar fashion, we treat the other sides and vertices. In order to write the linear system associated with the collocation problem, we construct a matrix, denoted by $\mathcal{A}_n$, that for $n = 4$ has the same structure as the one shown in Fig. 1.2.2. It turns out that $\mathcal{A}_n$ is the sum of the matrix $\mathcal{L}_n$ corresponding to the spectral discretization of the operator $-\Delta + \lambda I$, plus a matrix $\mathcal{C}_n$, taking care of the boundary constraints. Thus, recalling the definition of a tensor product given in (1.2.5), we have

$$(3.2.9) \quad \mathcal{A}_n := \mathcal{L}_n + \mathcal{C}_n = -[I_n \otimes (\tilde{D}_n^2 + \tfrac{1}{2}\lambda I_n) + (\tilde{D}_n^2 + \tfrac{1}{2}\lambda I_n) \otimes I_n]$$

$$+ [I_n \otimes K_n + K_n \otimes I_n],$$

where I_n is the $(n+1) \times (n+1)$ identity matrix, and K_n is defined by

$$(3.2.10) \quad K_n := \tfrac{1}{2}n(n+1) \begin{bmatrix} -\tilde{d}_{00}^{(1)} & -\tilde{d}_{01}^{(1)} & -\tilde{d}_{02}^{(1)} & \cdots & -\tilde{d}_{0n}^{(1)} \\ 0 & 0 & 0 & \cdots & 0 \\ 0 & 0 & 0 & \cdots & 0 \\ \vdots & \vdots & \vdots & \cdots & \vdots \\ 0 & 0 & 0 & \cdots & 0 \\ \tilde{d}_{n0}^{(1)} & \tilde{d}_{n1}^{(1)} & \tilde{d}_{n2}^{(1)} & \cdots & \tilde{d}_{nn}^{(1)} \end{bmatrix}$$

Denoting by $\vec{X}_n$ the unknown vector of the values of q_n at the Legendre nodes (1.2.3), we need to find the solution of the linear system (1.2.6). The right-hand side vector is now defined by $\vec{B}_n = \mathcal{J}_n^\Omega \vec{F}_n + \vec{H}_n$, where $\vec{F}_n$ is the vector of the values $f(\Theta_k^{(n)})$, $0 \leq k \leq n_T$, $\mathcal{J}_n^\Omega$ is the matrix defined in section 1.7, and $\vec{H}_n$ takes into account the values of the functions h_k, $1 \leq k \leq 4$, at the boundary nodes. For instance, when $n = 4$, we obtain:

$$\vec{H}_4 \equiv \tfrac{n(n+1)}{2}\Big([h_1(\eta_0^{(n)}) + h_4(\eta_0^{(n)})], h_1(\eta_1^{(n)}),$$

$$h_1(\eta_2^{(n)}), h_1(\eta_3^{(n)}), [h_1(\eta_4^{(n)}) + h_2(\eta_0^{(n)})],$$

$$h_4(\eta_1^{(n)}),\ 0\ ,\ 0\ ,\ 0\ , h_2(\eta_1^{(n)}),$$

$$h_4(\eta_2^{(n)}),\ 0\ ,\ 0\ ,\ 0\ , h_2(\eta_2^{(n)}),$$

$$h_4(\eta_3^{(n)}),\ 0\ ,\ 0\ ,\ 0\ , h_2(\eta_3^{(n)}),$$

$$[h_3(\eta_0^{(n)}) + h_4(\eta_4^{(n)})], h_3(\eta_1^{(n)}), h_3(\eta_2^{(n)}), h_3(\eta_3^{(n)}), [h_2(\eta_4^{(n)}) + h_3(\eta_4^{(n)})]\Big).$$

The proof that $\mathcal{A}_n$ is positive definite is straightforward, since the matrix is related, for $\lambda > 0$, to a positive bilinear form (take $\phi = q_n$ in (3.2.1)). Moreover, the eigenvalues are real because of the symmetry of the bilinear form (see also theorem 1.3.1). Hence, we can successfully apply an iterative method, such as the Du Fort-Frankel method, for the solution of the linear system $\mathcal{A}_n \vec{X}_n = \vec{B}_n$. To recover a fast convergence we must however introduce a preconditioner. This will be the goal of the next section.

3.3   The numerical algorithm for the Neumann problem

In order to construct a preconditioner for $\mathcal{A}_n$ we adopt a finite-differences scheme as done in section 1.4. As in (1.4.1), we use a five-points discretization, i.e.

$$(3.3.1) \qquad -\frac{2}{\eta_{j+1}^{(n)} - \eta_{j-1}^{(n)}}\left(\frac{z_{i,j+1} - z_{i,j}}{\eta_{j+1}^{(n)} - \eta_{j}^{(n)}} - \frac{z_{i,j} - z_{i,j-1}}{\eta_{j}^{(n)} - \eta_{j-1}^{(n)}}\right)$$

$$-\frac{2}{\eta_{i+1}^{(n)} - \eta_{i-1}^{(n)}}\left(\frac{z_{i+1,j} - z_{i,j}}{\eta_{i+1}^{(n)} - \eta_{i}^{(n)}} - \frac{z_{i,j} - z_{i-1,j}}{\eta_{i}^{(n)} - \eta_{i-1}^{(n)}}\right) + \lambda\, z_{i,j} = f(\eta_j^{(n)}, \eta_i^{(n)}),$$

for $1 \leq i \leq n - 1$, $1 \leq j \leq n - 1$. We now have to be careful when imposing the boundary conditions since we want to preserve the band structure. For example, at the side $]-1, 1[\times\{-1\}$, we may require for $1 \leq j \leq n - 1$:

$$(3.3.2) \qquad - \frac{2}{\eta_{j+1}^{(n)} - \eta_{j-1}^{(n)}} \left(\frac{z_{0,j+1} - z_{0,j}}{\eta_{j+1}^{(n)} - \eta_{j}^{(n)}} - \frac{z_{0,j} - z_{0,j-1}}{\eta_{j}^{(n)} - \eta_{j-1}^{(n)}} \right)$$

$$- \frac{1}{\tilde{w}_0^{(n)}} \left(\frac{z_{1,j} - z_{0,j}}{\eta_1^{(n)} - \eta_0^{(n)}} \right) + \lambda \, z_{0,j} = f(\eta_j^{(n)}, \eta_0^{(n)}) + \frac{1}{\tilde{w}_0^{(n)}} h_1(\eta_j^{(n)}).$$

At the side $\{1\} \times] - 1, 1[$, we may require for $1 \leq i \leq n - 1$:

$$(3.3.3) \qquad - \frac{2}{\eta_{i+1}^{(n)} - \eta_{i-1}^{(n)}} \left(\frac{z_{i+1,n} - z_{i,n}}{\eta_{i+1}^{(n)} - \eta_{i}^{(n)}} - \frac{z_{i,n} - z_{i-1,n}}{\eta_{i}^{(n)} - \eta_{i-1}^{(n)}} \right)$$

$$+ \frac{1}{\tilde{w}_n^{(n)}} \left(\frac{z_{i,n} - z_{i,n-1}}{\eta_n^{(n)} - \eta_{n-1}^{(n)}} \right) + \lambda \, z_{i,n} = f(\eta_n^{(n)}, \eta_i^{(n)}) + \frac{1}{\tilde{w}_n^{(n)}} h_2(\eta_i^{(n)}).$$

Similar equations will be imposed on the other two sides. We propose at $(-1, -1)$ the following equation

$$(3.3.4) \qquad - \frac{1}{\tilde{w}_0^{(n)}} \left(\frac{z_{1,0} - z_{0,0}}{\eta_1^{(n)} - \eta_0^{(n)}} \right) - \frac{1}{\tilde{w}_0^{(n)}} \left(\frac{z_{0,1} - z_{0,0}}{\eta_1^{(n)} - \eta_0^{(n)}} \right) + \lambda \, z_{0,0}$$

$$= f(\eta_0^{(n)}, \eta_0^{(n)}) + \frac{1}{\tilde{w}_0^{(n)}} h_1(\eta_0^{(n)}) + \frac{1}{\tilde{w}_0^{(n)}} h_4(\eta_0^{(n)}).$$

Something similar is done at the other corner points: $(1, -1)$, $(-1, 1)$, $(1, 1)$. As usual we denote by B_n the preconditioning matrix which has now the structure shown in Fig. 3.3.1 (compare with Fig. 1.4.1).

The behavior of the preconditioned eigenvalues is very good. For instance, if $\lambda = 1$, the eigenvalues of $B_n^{-1} A_n$ are real, positive and greater or equal to 1. The maximum preconditioned eigenvalue $\lambda_{\max}^{(n)}$ is about 2.82, mildly depending on n.

We apply the Du Fort-Frankel scheme, as explained in section 1.5, for the Poisson equation. The only drawback is that the value of $\mu^{(n)}$ estimated with (1.4.4) is not very close to $\lambda_{\max}^{(n)}$, so that the iterative method (1.4.6) slows down a bit. It is convenient to refine $\mu^{(n)}$ by a few more steps of the power method, i.e. we take for some $m \in \mathbf{N}$

$$(3.3.5) \qquad \mu^{(n)} := \frac{\|\vec{\Phi}_{n,m+1}\|}{\|\vec{\Phi}_{n,m}\|}, \qquad \text{where} \quad \vec{\Phi}_{n,m+1} := B_n^{-1} A_n \vec{\Phi}_{n,m},$$

and $\vec{\Phi}_{n,0} \equiv \vec{\Phi}_n$ is constructed as in section 1.4. With $m = 2$, we already get a fairly good approximation of $\lambda_{\max}^{(n)}$.

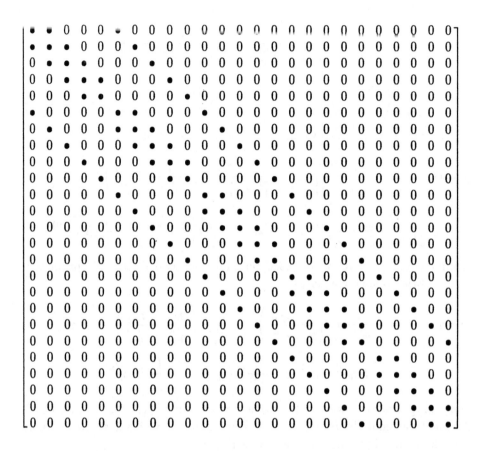

FIG. 3.3.1 - *Structure of the matrix B_4 with Neumann boundary conditions.*

For different n, we approximated the Neumann problem with $\lambda = 1$, $h_1 = h_2 = h_3 = h_4 = 0$ and $f(x, y) = xy$, $(x, y) \in \Omega$. The contour lines of the exact solution are given in Fig. 3.3.2. The norms of the residual, evaluated with (1.4.5), versus the number of iterations, are reported in Table 3.3.1. In the experiments we used $\mu^{(n)} = 2.82$ (although this value is not readily available) to show how fast is the convergence (and therefore the effectiveness of the preconditioner) with the optimal parameters. We remark that the residual is the sum of the error corresponding to the nodes inside Ω and the one corresponding to the boundary nodes.

At this point we can also consider the discretization of elliptic problems, where in part of the domain we assume Dirichlet-type conditions, while in the remaining part we impose Neumann boundary conditions.

FIG. 3.3.2 - *Solution to* $-\Delta U + U = xy$, *with homogeneous Neumann conditions.*

k	$n = 4$	$n = 8$	$n = 12$	$n = 16$	$n = 20$
2	$.569E - 1$	$.508E - 1$	$.337E - 1$	$.256E - 1$	$.207E - 1$
3	$.120E - 1$	$.170E - 1$	$.119E - 1$	$.919E - 2$	$.747E - 2$
4	$.118E - 2$	$.425E - 2$	$.334E - 2$	$.273E - 2$	$.229E - 2$
5	$.683E - 3$	$.106E - 2$	$.944E - 3$	$.805E - 3$	$.690E - 3$
6	$.262E - 3$	$.212E - 3$	$.233E - 3$	$.213E - 3$	$.190E - 3$
7	$.848E - 4$	$.367E - 4$	$.579E - 4$	$.568E - 4$	$.524E - 4$
8	$.136E - 4$	$.360E - 5$	$.131E - 4$	$.141E - 4$	$.136E - 4$
9	$.191E - 5$	$.696E - 6$	$.297E - 5$	$.354E - 5$	$.356E - 5$
10	$.744E - 6$	$.529E - 6$	$.595E - 6$	$.839E - 6$	$.887E - 6$

TABLE 3.3.1 - *Residuals of the approximation of the homogeneous Neumann problem.*

The approximating polynomial q_n will be required to satisfy the same kind of boundary constraints (in a weak form in the case of Neumann conditions) according to the position of each node in $\partial\Omega$. We examine the case of the equation (3.1.3) with $\lambda = 0$ and $f = 1$. We take homogeneous Neumann conditions at the side $]-1,1[\times\{-1\}$ and homogeneous Dirichlet conditions at the remaining sides and at the corner points. This boundary-value problem, which does not belong to the class covered by theorem 3.1.1, admits a unique solution. Therefore, we look for an approximated solution $q_n \in \mathbf{P}_n^*$ which satisfies (3.2.3) with $h_1 = 0$, and is vanishing at the other three sides of $\partial\Omega$. The contour lines of the exact solution are shown in Fig. 3.3.3 (compare with Fig. 1.5.1). Due to the smoothness of the solution, small degrees are sufficient in order to recover an accurate approximation. We give in Table 3.3.2 the norms of the corresponding residuals for different n (note that $\lambda_{max}^{(n)}$ has been approximated with $\mu^{(n)}$ obtained by (3.3.5) with $m = 2$).

k	$n = 4$	$n = 8$	$n = 12$	$n = 16$	$n = 20$
2	$.138E - 1$	$.112E - 1$	$.842E - 2$	$.614E - 2$	$.465E - 2$
3	$.213E - 2$	$.176E - 2$	$.175E - 2$	$.144E - 2$	$.116E - 2$
4	$.445E - 3$	$.271E - 3$	$.332E - 3$	$.312E - 3$	$.271E - 3$
5	$.321E - 4$	$.503E - 4$	$.599E - 4$	$.637E - 4$	$.596E - 4$
6	$.234E - 5$	$.106E - 4$	$.107E - 4$	$.125E - 4$	$.126E - 4$
7	$.287E - 6$	$.225E - 5$	$.202E - 5$	$.243E - 5$	$.262E - 5$
8	$.385E - 7$	$.532E - 6$	$.408E - 6$	$.471E - 6$	$.534E - 6$

TABLE 3.3.2 - *Residuals for the approximations of a Dirichlet-Neumann problem.*

One more example is obtained by imposing homogeneous Neumann conditions at the sides $]-1,1]\times\{-1\}$ and $\{1\}\times[-1,1[$, and homogeneous Dirichlet conditions at the remaining two sides (see the exact solution in Fig. 3.3.4). The numerical tests show a decay of the error in the iterative solution of the linear system which is similar to the one exhibited in the previous examples. We remark that in the construction of the preconditioning matrices we took into account the same boundary conditions required for the solution.

QUARTERONI and ZAMPIERI (1992) experimented other equivalent preconditioners based on finite elements.

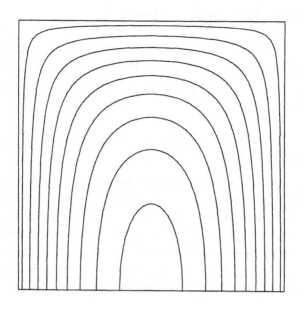

FIG. 3.3.3 - *Solution to $-\Delta U = 1$ with Neumann conditions at the side $]-1,1[\times\{-1\}$.*

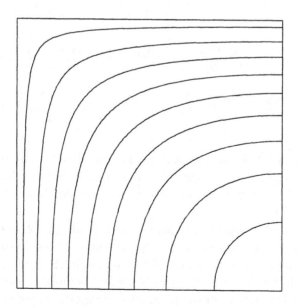

FIG. 3.3.4 - *Solution to $-\Delta U = 1$ with Neumann conditions at the sides $]-1,1]\times\{-1\}$ and $\{1\}\times[-1,1[$.*

3.4 More general boundary conditions

Extensions, concerning more general elliptic differential equations like (2.1.1) and more complicated conditions at the boundary, may be considered. For example, at the side $]-1,1[\times\{-1\}$, we can generalize the first equation in (3.1.1) by setting $\forall x \in]-1,1[$:

$$(3.4.1) \quad a_1(x)\frac{\partial U}{\partial x}(x,-1) + b_1(x)\frac{\partial U}{\partial y}(x,-1) + c_1(x)U(x,-1) = h_1(x),$$

where a_1, b_1, c_1 are given continuous functions. Similar conditions are imposed on the other sides. Of course, we shall assume that the coefficients are such that the new boundary-value problem admits a unique solution. We already examined in chapter two how to approximate the differential operator inside the domain by introducing a suitable upwind grid. We see how to handle the condition (3.4.1) in order to combine this with the upwind grid approach. A generalization of (3.2.3) leads to the equation

$$
(3.4.2) \qquad \left(f_1\frac{\partial^2 q_n}{\partial x^2} + f_2\frac{\partial^2 q_n}{\partial x\partial y} + f_3\frac{\partial^2 q_n}{\partial y^2} \right.
$$

$$
\left. + f_4\frac{\partial q_n}{\partial x} + f_5\frac{\partial q_n}{\partial y} + f_6\, q_n \right)(\eta_j^{(n)},\eta_0^{(n)})
$$

$$
+ \frac{1}{\tilde{w}_0^{(n)}}\left[a_1(\eta_j^{(n)})\frac{\partial q_n}{\partial x}(\eta_j^{(n)},\eta_0^{(n)}) + b_1(\eta_j^{(n)})\frac{\partial q_n}{\partial y}(\eta_j^{(n)},\eta_0^{(n)}) \right.
$$

$$
\left. + c_1(\eta_j^{(n)})q_n(\eta_j^{(n)},\eta_0^{(n)}) \right] = f_7(\eta_j^{(n)},\eta_0^{(n)}) + \frac{1}{\tilde{w}_0^{(n)}}h_1(\eta_j^{(n)}),
$$

for $1 \le j \le n-1$.

The condition (3.4.2) looks very natural and consistent, however, it may be difficult to interpret it as a byproduct of a positive bilinear form such as in (3.2.1). Therefore, there may be some troubles in developing a convergence theory for $n \to +\infty$.

In any event, the range of applications of formula (3.4.2) goes beyond the limit predicted by the theory, although we should always keep in mind the variational approach when possible. For example, in the case of the equation (2.1.4), the rule for the imposition of Neumann boundary conditions can be derived from a variational formulation, which suggests taking $a_1 = -\epsilon$, $b_1 = 0$ and $c_1 = 0$ in (3.4.1) and (3.4.2). By this approach the Neumann condition on $]-1,1[\times\{-1\}$ is imposed very weakly when ϵ is small. Nevertheless, by giving more weight to the normal derivative, setting for instance $a_1 = -1$ in

(3.4.2) (a choice that is not directly related to a variational formulation), we may find some instability troubles concerning the numerical implementation.

To build up the preconditioning matrix, we have to generalize (3.3.2). We take for simplicity $f_2 = 0$. We start by defining $\eta_{-1}^{(n)} := \eta_0^{(n)} - 2\tilde{w}_0^{(n)}$. Thus, for $1 \le j \le n-1$, we construct a finite-differences scheme, centered at $(\eta_j^{(n)}, \eta_0^{(n)})$ and based on the nine points $(\eta_{j+\beta}^{(n)}, \eta_\alpha^{(n)})$, where α and β belong to the set $\{-1, 0, 1\}$. Successively, we neglect the contributions related to the extra points $(\eta_{j+\beta}^{(n)}, \eta_{-1}^{(n)})$, $\beta \in \{-1, 0, 1\}$ (these quantities are actually zero if we assume that $f_3(\eta_j^{(n)}, \eta_0^{(n)}) = \tilde{w}_0^{(n)} f_5(\eta_j^{(n)}, \eta_0^{(n)}) + b_1(\eta_j^{(n)})$, as in the case of the equations (3.1.1)-(3.1.2)-(3.1.3)). This procedure yields

$$(3.4.3) \qquad f_1(\eta_j^{(n)}, \eta_0^{(n)}) \, \frac{2}{\eta_{j+1}^{(n)} - \eta_{j-1}^{(n)}} \left(\frac{z_{0,j+1} - z_{0,j}}{\eta_{j+1}^{(n)} - \eta_j^{(n)}} - \frac{z_{0,j} - z_{0,j-1}}{\eta_j^{(n)} - \eta_{j-1}^{(n)}} \right)$$

$$+ f_3(\eta_j^{(n)}, \eta_0^{(n)}) \, \frac{2}{\eta_1^{(n)} - \eta_{-1}^{(n)}} \left(\frac{z_{1,j} - z_{0,j}}{\eta_1^{(n)} - \eta_0^{(n)}} - \frac{z_{0,j}}{\eta_0^{(n)} - \eta_{-1}^{(n)}} \right)$$

$$+ \left[f_4(\eta_j^{(n)}, \eta_0^{(n)}) + \frac{1}{\tilde{w}_0^{(n)}} a_1(\eta_j^{(n)}) \right] \frac{z_{0,j+1} - z_{0,j-1}}{\eta_{j+1}^{(n)} - \eta_{j-1}^{(n)}}$$

$$+ \left[f_5(\eta_j^{(n)}, \eta_0^{(n)}) + \frac{1}{\tilde{w}_0^{(n)}} b_1(\eta_j^{(n)}) \right] \frac{z_{1,j}}{\eta_1^{(n)} - \eta_{-1}^{(n)}}$$

$$+ \left[f_6(\eta_j^{(n)}, \eta_0^{(n)}) + \frac{1}{\tilde{w}_0^{(n)}} c_1(\eta_j^{(n)}) \right] z_{0,j} = f_7(\eta_j^{(n)}, \eta_0^{(n)}) + \frac{1}{\tilde{w}_0^{(n)}} h_1(\eta_j^{(n)}).$$

The formula (3.3.2) is a special case of (3.4.3).

As far as the corner points are concerned, we follow a similar path. Regarding the node $(-1, -1)$, the reader can easily carry out the equations extending (3.2.7) and (3.3.4) to the general case.

In principle, the conditions to be imposed at the corners are suggested by the physics giving rise to the mathematical model, provided however that the differential problem is well-posed. In the case in which we do not have any specific prescription, and we want to construct some conditions at the corners compatible with the ones at the corresponding sides, the choice that we are going to present seems quite reasonable.

Let us suppose that at the side $]-1, 1[\times\{-1\}$ we have to satisfy (3.4.1) and at the side $\{1\}\times]-1, 1[$ we have $\forall y \in]-1, 1[$:

$$(3.4.4) \qquad a_2(y)\frac{\partial U}{\partial x}(1, y) + b_2(y)\frac{\partial U}{\partial y}(1, y) + c_2(y)U(1, y) = h_2(y).$$

Then at the point $(1, -1)$ we shall require that

(3.4.5) $$a_2(-1)\frac{\partial U}{\partial x}(1, -1) + b_1(1)\frac{\partial U}{\partial y}(1, -1)$$

$$+ [c_1(1) + c_2(-1)]U(1, -1) = h_1(1) + h_2(-1).$$

The same arguments hold true for the other corners. Note that in (3.4.5) the terms associated with the tangential derivatives $a_1(x)\frac{\partial U}{\partial x}(x, -1)$ and $b_2(x)\frac{\partial U}{\partial y}(1, y)$ have been eliminated.

We see two examples where the condition imposed at $(1, -1)$ respects (3.4.5). In the first one, we look for U such that

(3.4.6)
$$
\begin{cases}
-\Delta U = 1 & \text{in } \Omega =]-1, 1[\times]-1, 1[, \\[2mm]
\left(\dfrac{\partial U}{\partial x} - \dfrac{\partial U}{\partial y}\right)(x, -1) = 0 & \forall x \in]-1, 1[, \\[2mm]
\left(\dfrac{\partial U}{\partial x} - \dfrac{\partial U}{\partial y}\right)(1, y) = 0 & \forall y \in]-1, 1[, \\[2mm]
\left(\dfrac{\partial U}{\partial x} - \dfrac{\partial U}{\partial y}\right)(1, -1) = 0 \\[2mm]
U(x, 1) = 0 & \forall x \in [-1, 1], \\[2mm]
U(-1, y) = 0 & \forall y \in [-1, 1].
\end{cases}
$$

The contour lines of U are shown in Fig. 3.4.1. In the second problem, we look for V satisfying:

(3.4.7)
$$
\begin{cases}
-\Delta V = 1 & \text{in } \Omega =]-1, 1[\times]-1, 1[, \\[2mm]
\left(-\dfrac{\partial V}{\partial x} - \dfrac{\partial V}{\partial y}\right)(x, -1) = 0 & \forall x \in]-1, 1[, \\[2mm]
\left(\dfrac{\partial V}{\partial x} + \dfrac{\partial V}{\partial y}\right)(1, y) = 0 & \forall y \in]-1, 1[, \\[2mm]
\left(\dfrac{\partial V}{\partial x} - \dfrac{\partial V}{\partial y}\right)(1, -1) = 0 \\[2mm]
V(x, 1) = 0 & \forall x \in [-1, 1], \\[2mm]
V(-1, y) = 0 & \forall y \in [-1, 1].
\end{cases}
$$

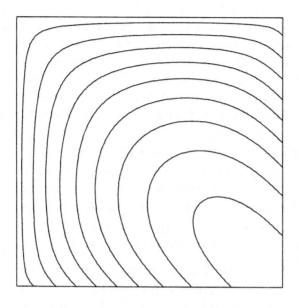

FIG. 3.4.1 - *Solution to* (3.4.6).

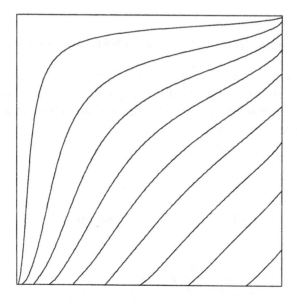

FIG. 3.4.2 - *Approximated solution to* (3.4.7) *for* $n = 12$.

The preconditioned Du Fort-Frankel method, applied to the linear system obtained by collocating the boundary-value problems (3.4.6) and (3.4.7), allows us to compute the approximated solutions. In particular, in the case of problem (3.4.6), we get a fast convergence. For instance, when $n = 12$ and $m = 2$ in (3.3.5), we obtain in 20 iterations an error on the residual of the order of 10^{-6}. The collocation problem associated with (3.4.7) is a bit nastier from the numerical point of view. The convergence is however reached in a reasonable amount of iterations. We provide in Fig. 3.4.2 the plot of the approximating polynomial of degree $n = 12$.

It is worthwhile to note that the condition $2 \left(\frac{\partial}{\partial x} U - \frac{\partial}{\partial y} U \right) = 0$ at $(1, -1)$, obtained by summing up the boundary relations at the sides $] - 1, 1[\times \{-1\}$ and $\{1\} \times] - 1, 1[$, is algebraically equivalent to the one imposed in (3.4.6), but the rate of convergence of the iterative method is lower due to a loss of efficiency of the preconditioner. In the same way, summing up the boundary relations in (3.4.7), we would obtain the condition $\left(-\frac{\partial}{\partial x} V - \frac{\partial}{\partial y} V \right) (1, -1) + \left(\frac{\partial}{\partial x} V + \frac{\partial}{\partial y} V \right) (1, -1) = 0$. This brings catastrophic consequences since the determinant of the linear system corresponding to the collocation problem turns out to be zero. We further support (3.4.5) by observing that the condition at $(-1, 1)$ given in (3.4.7) agrees with the fact that V is symmetric with respect to the straight line $x = -y$.

3.5 An approximation of the Poincaré-Steklov operator

We give here some results which will be useful later on. We define $\Gamma :=$ $] - 1, 1[\times \{-1\}$ and we study the following question: given the function $h :] - 1, 1[\rightarrow \mathbf{R}$, find the function $g :] - 1, 1[\rightarrow \mathbf{R}$, such that the solution $U : \bar{\Omega} \rightarrow \mathbf{R}$ to the boundary-value problem

$$(3.5.1) \qquad \begin{cases} -\Delta U = 0 & \text{in } \Omega, \\[2mm] U = g & \text{on } \Gamma, \\[2mm] U = 0 & \text{on } \partial\Omega - \Gamma, \end{cases}$$

satisfies

$$(3.5.2) \qquad h(x) = -\frac{\partial U}{\partial y}(x, -1) \qquad \forall x \in] - 1, 1[.$$

In other words, the function g has to be determined in such a way that the solution U of (3.5.1) has the outward normal derivative on Γ equal to h. Assuming some regularity for the function h, we can determine a unique solution

g to our problem. Thus, we can define a *linear* application π, known as the *Poincaré-Steklov operator*, such that $\pi(h) = g$. A more rigorous introduction to the Poincaré-Steklov operator, some theoretical remarks, and a list of references are provided in AGOSHKOV (1988).

Let us now work in the space of polynomials $\mathbf{P}_n^\star$, $n \geq 2$, to find a discrete counterpart of the operator π. We start by assigning a continuous function $g :]-1, 1[\to \mathbf{R}$ and construct the vector $\vec{g}_n := \{g(\eta_j^{(n)})\}_{1 \leq j \leq n-1}$. Successively, we find the solution $q_n \in \mathbf{P}_n^\star$ to the collocation problem

$$(3.5.3) \quad \begin{cases} (-\Delta q_n)(\eta_j^{(n)}, \eta_i^{(n)}) = 0 & 1 \leq j \leq n-1, \ 1 \leq i \leq n-1, \\[2mm] q_n(\eta_j^{(n)}, \eta_0^{(n)}) = g(\eta_j^{(n)}) & 1 \leq j \leq n-1, \\[2mm] q_n = 0 & \text{at the other boundary nodes.} \end{cases}$$

Finally, we consider the vector

$$(3.5.4) \quad \vec{h}_n := \left\{ (-\Delta q_n)(\eta_j^{(n)}, \eta_0^{(n)}) \, \tilde{w}_0^{(n)} - \frac{\partial q_n}{\partial y}(\eta_j^{(n)}, \eta_0^{(n)}) \right\}_{1 \leq j \leq n-1}$$

which represents the normal derivative to Γ in the weak sense (see section 3.2).

Therefore, we can define a linear transformation, corresponding to a $(n-1) \times (n-1)$ matrix $\mathcal{P}_n$, such that

$$(3.5.5) \quad \vec{h}_n = \mathcal{P}_n \vec{g}_n.$$

It is not difficult to verify that $\mathcal{P}_n$ is invertible (if $\vec{h}_n \equiv 0$, then we use (3.2.1) to obtain $q_n = 0$, which in turn yields $\vec{g}_n \equiv 0$) and we define $\mathcal{P}_n^{-1}$ to be the discrete version of the Poincaré-Steklov operator. To determine the entries of $\mathcal{P}_n$, we consider the set of Lagrange polynomials $\tilde{l}_k^{(n)}$, $0 \leq k \leq n$, with respect to the Legendre nodes (see $(A.3.1)$ and $(A.3.2)$). The vectors $\vec{h}_n$, obtained by taking $g = \tilde{l}_k^{(n)}$, for $1 \leq k \leq n-1$, respectively, are the columns of the matrix $\mathcal{P}_n$, which turns out to be full. In this way, the procedure for the construction of $\mathcal{P}_n$ is quite expensive since it amounts to solving a set of $n-1$ collocation problems. Consequently, our new goal is to propose another $(n-1) \times (n-1)$ matrix $\tilde{\mathcal{P}}_n$, which is a suitable approximation of $\mathcal{P}_n$ and can be computed in a very inexpensive way. To this end, we denote by $\zeta_{i,j}$, $1 \leq i \leq n-1$, $1 \leq j \leq n-1$, the entries of $\mathcal{P}_n$, and we recall that $\xi_k^{(n)}$, $1 \leq k \leq n$, are the zeroes of the Legendre polynomial P_n (see section A.2). By carrying out some numerical tests we found out the following estimates

$$(3.5.6) \quad \zeta_{i,i} \approx \frac{5}{3(\xi_{i+1}^{(n)} - \xi_i^{(n)})} \qquad 1 \leq i \leq n-1,$$

(3.5.7)
$$\zeta_{i,i-1} \approx \frac{-6}{6(\xi_{i+1}^{(n)} - \eta_{i-1}^{(n)})} \qquad 2 \le i \le n-1,$$

(3.5.8)
$$\zeta_{i,i+1} \approx \frac{-5}{6(\eta_{i+1}^{(n)} - \xi_i^{(n)})} \qquad 1 \le i \le n-2,$$

(3.5.9)
$$\zeta_{i,j} \approx 0 \qquad \text{for all other entries.}$$

Therefore, when $g = \tilde{l}_i^{(n)}$ in (3.5.3), only three entries (the ones relative to the indices $j = i-1$, $j = i$, $j = i+1$) are significantly different from zero, moreover they are very well approximated by (3.5.6)-(3.5.7)-(3.5.8). This allows us to substitute $\mathcal{P}_n$ by a tridiagonal non-symmetric matrix $\tilde{\mathcal{P}}_n$, which is diagonally dominant. In section 4.4, $\tilde{\mathcal{P}}_n^{-1}$ will be used as an approximation of the Poincaré-Steklov operator.

We finally note that, if $q_n \in \mathbf{P}_n^*$ satisfies the collocation problem

(3.5.10)
$$\begin{cases} (-\Delta q_n)(\eta_j^{(n)}, \eta_i^{(n)}) = 0 & 1 \le j \le n-1, \ 1 \le i \le n-1, \\ q_n(\eta_0^{(n)}, \eta_0^{(n)}) = 1, \\ q_n = 0 & \text{at the other boundary nodes,} \end{cases}$$

then the explicit solution is: $q_n(x,y) = \tilde{l}_0^{(n)}(x)\tilde{l}_0^{(n)}(y)$, $(x,y) \in \bar{\Omega}$. A simple computation shows that

(3.5.11) $\left(-\Delta q_n \ \tilde{w}_0^{(n)} - \dfrac{\partial q_n}{\partial x} - \dfrac{\partial q_n}{\partial y} \right)(\eta_0^{(n)}, \eta_0^{(n)}) = \frac{1}{3}[n(n+1) + 1].$

The situation at the other corners of $\bar{\Omega}$ is very similar. This result will also be used in section 4.4.

4

The Spectral Element
Method

We analyze the discretization of elliptic boundary-value problems defined in domains with a complicated shape, via a domain decomposition approach. The approximated solution is a patchwork of different algebraic polynomials defined in the subdomains and is determined as the result of a preconditioned iterative procedure. At any iteration, in any subdomain, the corresponding polynomial satisfies a collocation problem.

4.1 Complex geometries

We are concerned with finding approximated solutions to boundary-value problems like the ones studied in the previous chapters. They will be defined in an open domain Ω, which is more complex than a square. We shall require that $\bar{\Omega} = \cup_{m=1}^{M}\bar{\Omega}_m$ is the union of a finite number M of disjoint convex quadrilateral regions Ω_m. Hence, $\partial\Omega_m$ has four *sides* and four *vertices*. From now on we assume that each side does not include the endpoints. We shall also make the assumption that this decomposition of Ω is of *conforming* type, which means that, for any $1 \leq m \leq M$, the vertices of $\bar{\Omega}_m$ are not allowed to fall within the side of another subdomain of the partition.

An example of domain Ω is the "house", shown in Figs. 4.1.1, 4.1.2 and 4.1.3, which has been decomposed in a conforming way into different numbers of subdomains. The unity of measure is provided at the top left of Fig. 4.1.1. We shall often deal in the future with this domain since it displays some interesting peculiarities. For instance, because of the "window", Ω is not a simply-connected region. Moreover, it has some "peninsulas" (such as the "chimney") and many *cross-points*, i.e. points not belonging to the boundary $\partial\Omega$ at the meeting of vertices associated with different subdomains. There are four cross-points in Fig. 4.1.1, 25 in Fig. 4.1.2, and 58 in Fig. 4.1.3.

A cross-point can be at the intersection of three domains ($3d$-type cross-point) or four domains ($4d$-type cross-point), but for simplicity no more than four subdomains will be allowed to meet at one point. In the decomposition of the set of Fig. 4.1.4 (the "diamond") there are three cross-points: two of which are $3d$-type and one is $4d$-type.

The term *interface* will generically denote the parts inside Ω shared by two or more subdomains. Sides in common or cross-points are interfaces. A boundary point at the intersection of two subdomains is not an interface point.

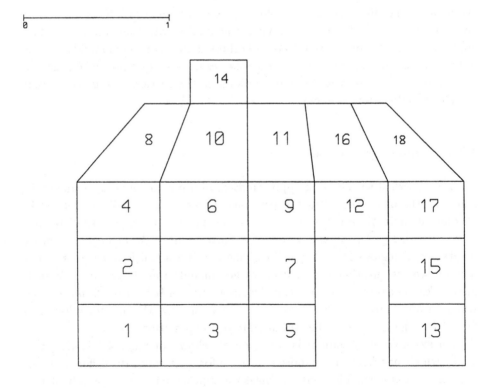

FIG. 4.1.1 - *The "house" decomposed in 18 subdomains.*

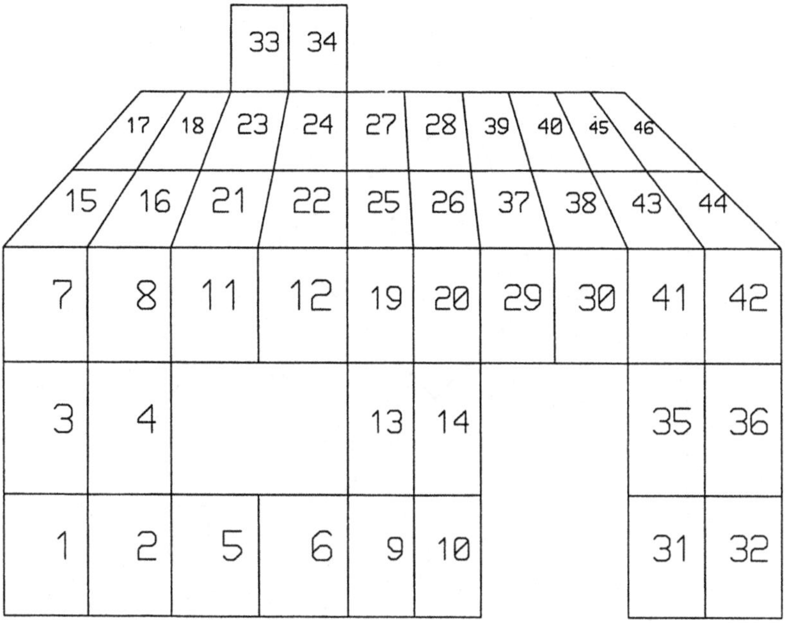

FIG. 4.1.2 - The "house" decomposed in 46 subdomains.

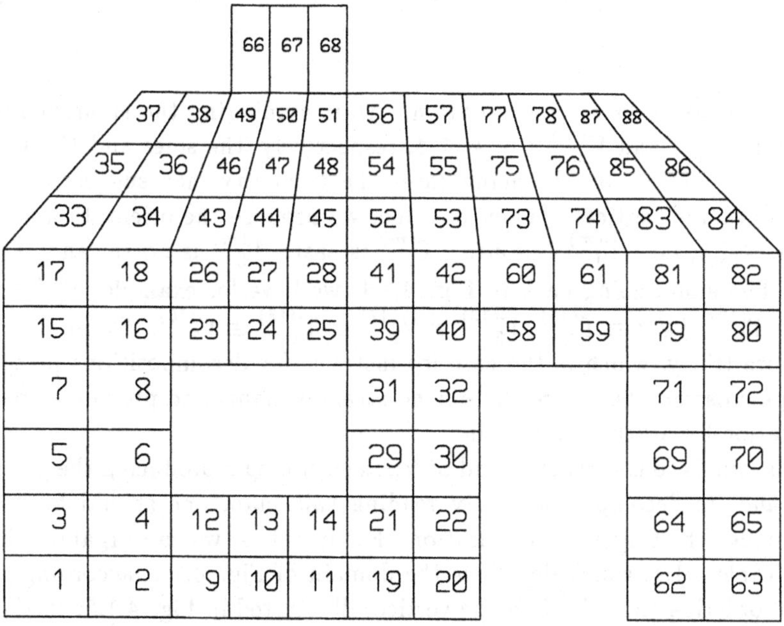

FIG. 4.1.3 - The "house" decomposed in 88 subdomains.

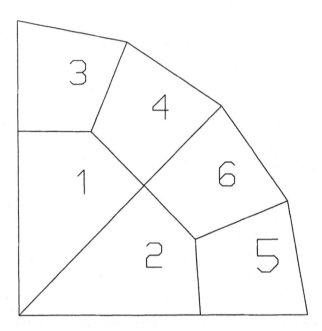

FIG. 4.1.4 - *Decomposition of the "diamond" with cross-points of different type.*

We now introduce some notation. After numbering the subdomains, we denote by $\Gamma_k^{(m)}$ and $V_k^{(m)}$ $1 \le k \le 4$, respectively, the sides and the vertices of $\bar{\Omega}_m$, $1 \le m \le M$. In many cases the geometry may suggest a way to order sides and vertices. When possible, we propose the orientation adopted in Fig. 4.1.5, i.e. $\Gamma_1^{(m)}$ is south, $\Gamma_2^{(m)}$ is east, $\Gamma_3^{(m)}$ is north, and $\Gamma_4^{(m)}$ is west. Therefore, going back to Fig. 4.1.1, we have for example: $\Gamma_3^{(1)} \equiv \Gamma_1^{(2)}$, $\Gamma_3^{(6)} \equiv \Gamma_1^{(10)}$, $\Gamma_2^{(11)} \equiv \Gamma_4^{(16)}$, $V_3^{(12)} \equiv V_4^{(17)} \equiv V_1^{(18)} \equiv V_2^{(16)}$. Other situations are more tricky, which is the case for instance for decompositions presenting $3d$-type cross-points, where there is no straight manner to provide a common orientation for all the subdomains.

Our aim is to construct a kind of *topological map* reproducing the geometry of Ω and its decomposition. By reading this map, one should be able to reconstruct the original configuration. For instance, we can organize the job as in Table 4.1.1, which describes the domain of Fig. 4.1.1, according to the way of ordering the sides and the vertices illustrated in Fig. 4.1.5. In the first column, we report the coordinates of the vertices. For any $V_k^{(m)}$, in column A, we specify the number of the subdomain next to $\bar{\Omega}_m$ sharing with it side $\Gamma_k^{(m)}$.

The number of the side matching with $\Gamma_k^{(m)}$ and belonging to the neighboring subdomain is reported in column B. The entry is set to 0 if $\Gamma_k^{(m)}$ is a boundary side. If $V_k^{(m)}$ is a 4d-type cross-point, then column C gives the number of the domain sharing with $\bar{\Omega}_m$ only the single point $V_k^{(m)}$. The entry is set to 0 if $V_k^{(m)}$ is a boundary vertex.

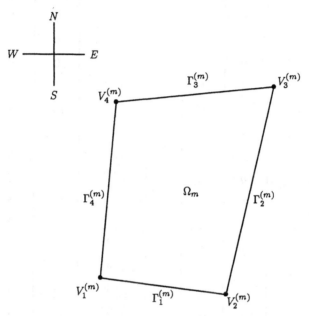

FIG. 4.1.5 - *Sides and vertices of the domain* $\bar{\Omega}_m$.

In order to use the results developed in the previous chapters, we will transform every $\bar{\Omega}_m$, $1 \leq m \leq M$, into a *reference* square $S =\ ]-1, 1[\times]-1, 1[$. Thus, for any $1 \leq m \leq M$, we define an *isoparametric* mapping $\theta_m : \bar{S} \to \bar{\Omega}_m$ which brings the points $(-1, -1)$, $(1, -1)$, $(1, 1)$, $(-1, 1)$ into the points $V_1^{(m)}$, $V_2^{(m)}$, $V_3^{(m)}$, $V_4^{(m)}$, respectively (see Fig. 4.1.6). Denoting by (x, y) the coordinates of a point in $\bar{\Omega}_m$ and by $(\hat{x}, \hat{y})$ the coordinates of a point in $\bar{S}$, the two components of the transformation $\theta_m \equiv (\theta_{1,m}, \theta_{2,m})$ are expressed by the following formulas:

$$(4.1.1) \quad x \ = \ \theta_{1,m}(\hat{x}, \hat{y}) \ = \ \tfrac{1}{4}\Big[(a_1 - a_2 + a_3 - a_4)\hat{x}\hat{y} \ + \ (-a_1 + a_2 + a_3 - a_4)\hat{x}$$

$$+ \ (-a_1 - a_2 + a_3 + a_4)\hat{y} \ + \ (a_1 + a_2 + a_3 + a_4)\Big],$$

vertex	A	B	C
$V_1^{(1)}$	0	0	0
$V_2^{(1)}$	3	4	0
$V_3^{(1)}$	2	1	0
$V_4^{(1)}$	0	0	0
$V_1^{(2)}$	1	3	0
$V_2^{(2)}$	0	0	0
$V_3^{(2)}$	4	1	0
$V_4^{(2)}$	0	0	0
$V_1^{(3)}$	0	0	0
$V_2^{(3)}$	5	4	0
$V_3^{(3)}$	0	0	0
$V_4^{(3)}$	1	2	0
$V_1^{(4)}$	2	3	0
$V_2^{(4)}$	6	4	0
$V_3^{(4)}$	8	1	10
$V_4^{(4)}$	0	0	0
$V_1^{(5)}$	0	0	0
$V_2^{(5)}$	0	0	0
$V_3^{(5)}$	7	1	0
$V_4^{(5)}$	3	2	0
$V_1^{(6)}$	0	0	0
$V_2^{(6)}$	9	4	0
$V_3^{(6)}$	10	1	11
$V_4^{(6)}$	4	2	8

vertex	A	B	C
$V_1^{(7)}$	5	3	0
$V_2^{(7)}$	0	0	0
$V_3^{(7)}$	9	1	0
$V_4^{(7)}$	0	0	0
$V_1^{(8)}$	4	3	0
$V_2^{(8)}$	10	4	6
$V_3^{(8)}$	0	0	0
$V_4^{(8)}$	0	0	0
$V_1^{(9)}$	7	3	0
$V_2^{(9)}$	12	4	0
$V_3^{(9)}$	11	1	16
$V_4^{(9)}$	6	2	10
$V_1^{(10)}$	6	3	4
$V_2^{(10)}$	11	4	9
$V_3^{(10)}$	14	1	0
$V_4^{(10)}$	8	2	0
$V_1^{(11)}$	9	3	6
$V_2^{(11)}$	16	4	12
$V_3^{(11)}$	0	0	0
$V_4^{(11)}$	10	2	0
$V_1^{(12)}$	0	0	0
$V_2^{(12)}$	17	4	0
$V_3^{(12)}$	16	1	18
$V_4^{(12)}$	9	2	11

vertex	A	B	C
$V_1^{(13)}$	0	0	0
$V_2^{(13)}$	0	0	0
$V_3^{(13)}$	15	1	0
$V_4^{(13)}$	0	0	0
$V_1^{(14)}$	10	3	0
$V_2^{(14)}$	0	0	0
$V_3^{(14)}$	0	0	0
$V_4^{(14)}$	0	0	0
$V_1^{(15)}$	13	3	0
$V_2^{(15)}$	0	0	0
$V_3^{(15)}$	17	1	0
$V_4^{(15)}$	0	0	0

vertex	A	B	C
$V_1^{(16)}$	12	3	9
$V_2^{(16)}$	18	4	17
$V_3^{(16)}$	0	0	0
$V_4^{(16)}$	11	2	0
$V_1^{(17)}$	15	3	0
$V_2^{(17)}$	0	0	0
$V_3^{(17)}$	18	1	0
$V_4^{(17)}$	12	2	16
$V_1^{(18)}$	17	3	12
$V_2^{(18)}$	0	0	0
$V_3^{(18)}$	0	0	0
$V_4^{(18)}$	16	2	0

TABLE 4.1.1 - *Topological map of the "house" decomposed in 18 subdomains.*

$$(4.1.2) \quad y = \theta_{2,m}(\hat{x}, \hat{y}) = \tfrac{1}{4}\Big[(b_1 - b_2 + b_3 - b_4)\hat{x}\hat{y} + (-b_1 + b_2 + b_3 - b_4)\hat{x}$$

$$+ (-b_1 - b_2 + b_3 + b_4)\hat{y} + (b_1 + b_2 + b_3 + b_4)\Big].$$

Note that θ_m is invertible.

A function U defined in Ω_m can also be represented through the function $\hat{U} = U(\theta_m)$ defined in S. The derivatives of $\hat{U}$ are computed by the usual rules:

$$(4.1.3) \quad \frac{\partial \hat{U}}{\partial \hat{x}} = \frac{\partial U}{\partial x}\frac{\partial \theta_{1,m}}{\partial \hat{x}} + \frac{\partial U}{\partial y}\frac{\partial \theta_{2,m}}{\partial \hat{x}}, \qquad \frac{\partial \hat{U}}{\partial \hat{y}} = \frac{\partial U}{\partial x}\frac{\partial \theta_{1,m}}{\partial \hat{y}} + \frac{\partial U}{\partial y}\frac{\partial \theta_{2,m}}{\partial \hat{y}}.$$

High order derivatives are evaluated in a very similar way.

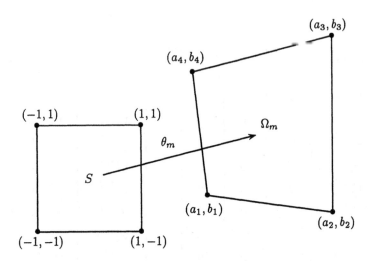

FIG. 4.1.6 - *The isoparametric map θ_m.*

The approximation technique which we are presently going to examine is based on the variational approach of the finite element method in order to deal with the decomposition in subdomains, while techniques typical of spectral methods are used to handle the solution inside the various pieces. The result is the *spectral element* method. Although the adoption of spectral elements was considered in GOTTLIEB and ORSZAG (1977), ORSZAG (1980), METIVET and MORCHOISNE (1982), and MORCHOISNE (1983), the terminology was not introduced until PATERA (1984). Successive extensions and applications are due to many authors. Most of spectral element papers are specifically related to the discretization of the Navier-Stokes equations (see also sections 5.2 and 5.3). A list of all references in domain decomposition techniques for spectral methods would be certainly too long. We give here only various contributions: HALDEN-WANG (1984), ZANG and HUSSAINI (1985), FUNARO (1986), MACARAEG and STREETT (1986), KOPRIVA (1986), KORCZAK and PATERA (1986), CANUTO and FUNARO (1988), FUNARO, QUARTERONI and ZANOLLI (1988), LACROIX, PEYRET and PULICANI (1988), PULICANI (1988), QUARTERONI and SACCHI LANDRIANI (1988), RØNQUIST (1988), DEMARET, DEVILLE and SCHNEIDESCH (1989), EHRENSTEIN, GUILLARD and PEYRET (1989), KARNIADAKIS (1989), KU, HIRSH, TAYLOR and ROSEMBERG (1989), PHILLIPS and KARAGEORGHIS (1989), FARCY and ALZIARY DE ROQUEFORT (1990), PEYRET (1990), ISRAELI, VOZOVOI and AVERBUCH (1993), FISHER and RØNQUIST (1994), and BERNAR-DI, MADAY and PATERA (1997).

4.2 The Poisson equation in a complex domain

We study here, and in sections 4.3 and 4.4, the equation (1.1.1) where Ω is a domain of the type previously considered. Later, in section 4.6, we will examine transport-diffusion equations. For simplicity, we assume homogeneous Dirichlet boundary conditions, i.e. $U = 0$ on $\partial\Omega$, although the case of non vanishing conditions can be treated in the same manner. The Poisson problem on Ω admits the same variational formulation (1.1.6). The function U is supposed to be continuous and, having $\bar\Omega = \cup_{m=1}^{M}\bar\Omega_m$, we define U_m to be the restriction of U to the set $\bar\Omega_m$. Then, (1.1.6) yields

$$(4.2.1) \qquad B(U,\phi) = \sum_{m=1}^{M}\int_{\Omega_m}(\vec\nabla U_m\cdot\vec\nabla\phi)dxdy$$

$$= \sum_{m=1}^{M}\int_{\partial\Omega_m}\frac{\partial U_m}{\partial\vec\nu^{(m)}}\phi\,ds - \sum_{m=1}^{M}\int_{\Omega_m}\Delta U_m\phi\,dxdy$$

$$= \sum_{m=1}^{M}\int_{\Omega_m}f\phi\,dxdy = F(\phi),$$

with $\vec\nu^{(m)}$, $1 \le m \le M$, denoting the unitary outward normal vector to the boundary of Ω_m. We recall that U_m is required to vanish on $\partial\Omega_m \cap \partial\Omega$. By testing on functions ϕ vanishing on $\Omega - \Omega_m$ we get

$$(4.2.2) \qquad -\Delta U_m = f \quad \text{in } \Omega_m, \quad 1 \le m \le M.$$

If Ω_r and Ω_s, $r \ne s$, are two domains having a common side Γ, we set $\vec\nu := \vec\nu^{(r)} = -\vec\nu^{(s)}$ on Γ, and by (4.2.1) we can also deduce

$$(4.2.3) \qquad \frac{\partial}{\partial\vec\nu}U_r = \frac{\partial}{\partial\vec\nu}U_s \quad \text{on } \Gamma.$$

The relation (4.2.3) says that the function $\frac{\partial}{\partial\vec\nu}U$ is continuous on Γ, or, in other words, that there is no jump of the normal derivatives across Γ.

We now use the mapping θ_m, $1 \le m \le M$, defined in section 4.1 and we evaluate the derivatives as in (4.1.3). We also note that: $\frac{\partial^2}{\partial\hat{x}^2}\theta_{k,m} = \frac{\partial^2}{\partial\hat{y}^2}\theta_{k,m} = 0$, $1 \le k \le 2$. Then, as with (2.1.1), one writes the equation (4.2.2) in the reference square S, i.e.

$$(4.2.4) \qquad L_m\hat{U}_m := f_{1,m}\frac{\partial^2\hat{U}_m}{\partial\hat{x}^2} + f_{2,m}\frac{\partial^2\hat{U}_m}{\partial\hat{x}\partial\hat{y}} + f_{3,m}\frac{\partial^2\hat{U}_m}{\partial\hat{y}^2}$$

$$+ f_{4,m}\frac{\partial\hat{U}_m}{\partial\hat{x}} + f_{5,m}\frac{\partial\hat{U}_m}{\partial\hat{y}} + f_{6,m}\hat{U}_m = f_{7,m},$$

where the coefficients of the operator L_m are

$$(4.2.5) \qquad f_{1,m} := -\frac{1}{\sigma_m^2}\left[\left(\frac{\partial\theta_{1,m}}{\partial\hat{y}}\right)^2 + \left(\frac{\partial\theta_{2,m}}{\partial\hat{y}}\right)^2\right],$$

$$(4.2.6) \qquad f_{2,m} := \frac{2}{\sigma_m^2}\left[\frac{\partial\theta_{1,m}}{\partial\hat{x}}\frac{\partial\theta_{1,m}}{\partial\hat{y}} + \frac{\partial\theta_{2,m}}{\partial\hat{x}}\frac{\partial\theta_{2,m}}{\partial\hat{y}}\right],$$

$$(4.2.7) \qquad f_{3,m} := -\frac{1}{\sigma_m^2}\left[\left(\frac{\partial\theta_{1,m}}{\partial\hat{x}}\right)^2 + \left(\frac{\partial\theta_{2,m}}{\partial\hat{x}}\right)^2\right],$$

$$(4.2.8) \qquad f_{4,m} := \frac{f_{2,m}}{\sigma_m}\left[\frac{\partial\theta_{1,m}}{\partial\hat{y}}\frac{\partial^2\theta_{2,m}}{\partial\hat{x}\partial\hat{y}} - \frac{\partial\theta_{2,m}}{\partial\hat{y}}\frac{\partial^2\theta_{1,m}}{\partial\hat{x}\partial\hat{y}}\right],$$

$$(4.2.9) \qquad f_{5,m} := \frac{f_{2,m}}{\sigma_m}\left[\frac{\partial\theta_{2,m}}{\partial\hat{x}}\frac{\partial^2\theta_{1,m}}{\partial\hat{x}\partial\hat{y}} - \frac{\partial\theta_{1,m}}{\partial\hat{x}}\frac{\partial^2\theta_{2,m}}{\partial\hat{x}\partial\hat{y}}\right],$$

$$(4.2.10) \qquad f_{6,m} := 0, \qquad f_{7,m} := f(\theta_m),$$

and

$$(4.2.11) \qquad \sigma_m := \frac{\partial\theta_{1,m}}{\partial\hat{x}}\frac{\partial\theta_{2,m}}{\partial\hat{y}} - \frac{\partial\theta_{1,m}}{\partial\hat{y}}\frac{\partial\theta_{2,m}}{\partial\hat{x}}.$$

The function $\sigma_m : \bar{S} \to \mathbf{R}$ is the determinant of the Jacobian J_m of θ_m and vanishes only if the measure of Ω_m is zero, which is a situation certainly to be avoided. Moreover, the functions $f_{2,m}$, $f_{4,m}$, and $f_{5,m}$ are vanishing when the transformation θ_m is *angle preserving*, i.e. if Ω_m is a rectangle. We obtain an explicit expression of the coefficients $f_{k,m}$, $1 \le k \le 7$, with the help of (4.1.1) and (4.1.2).

Let us see what happens to the relation (4.2.3) when using the mappings θ_r and θ_s. With the notation introduced in section 4.1, we analyze for the case when $\Gamma \equiv \Gamma_2^{(r)} \equiv \Gamma_4^{(s)}$ (the same arguments hold true for the case $\Gamma \equiv \Gamma_1^{(r)} \equiv \Gamma_3^{(s)}$). We recall that both $\hat{U}_r$ and $\hat{U}_s$ are defined on $\bar{S}$. Then from (4.1.3), for $\hat{y} \in\,]-1, 1[$, we have

$$(4.2.12) \qquad \frac{1}{\sigma_r}\left[\left(\nu_1\frac{\partial\theta_{2,r}}{\partial\hat{y}} - \nu_2\frac{\partial\theta_{1,r}}{\partial\hat{y}}\right)\frac{\partial\hat{U}_r}{\partial\hat{x}}\right.$$

$$\left. + \left(-\nu_1\frac{\partial\theta_{2,r}}{\partial\hat{x}} + \nu_2\frac{\partial\theta_{1,r}}{\partial\hat{x}}\right)\frac{\partial\hat{U}_r}{\partial\hat{y}}\right](1,\hat{y}) =$$

$$= \frac{1}{\sigma_s} \left[\left(\nu_1 \frac{\partial \theta_{2,s}}{\partial \hat{y}} - \nu_2 \frac{\partial \theta_{1,s}}{\partial \hat{y}} \right) \frac{\partial \hat{U}_s}{\partial \hat{x}} \right.$$

$$\left. + \left(-\nu_1 \frac{\partial \theta_{2,s}}{\partial \hat{x}} + \nu_2 \frac{\partial \theta_{1,s}}{\partial \hat{x}} \right) \frac{\partial \hat{U}_s}{\partial \hat{y}} \right] (-1, \hat{y}),$$

since we used the fact that $\frac{\partial}{\partial \vec{\nu}} U = \nu_1 \frac{\partial}{\partial x} U + \nu_2 \frac{\partial}{\partial y} U$ with $\vec{\nu} \equiv (\nu_1, \nu_2)$ and $\nu_1^2 + \nu_2^2 = 1$. This shows that $\frac{\partial}{\partial \vec{\mu}_r} \hat{U}_r$ on $\{1\} \times] - 1, 1[$ is equal to $\frac{\partial}{\partial \vec{\mu}_s} \hat{U}_s$ on $\{-1\} \times] - 1, 1[$, where $\vec{\mu}_r$ and $\vec{\mu}_s$ are suitable vector fields defined on ∂S.

Another way to get (4.2.4) and (4.2.12) is to rewrite (4.2.1) by changing the variables of integration in each domain. This implies

(4.2.13)
$$\sum_{m=1}^{M} \int_S [({}^t J_m^{-1} \vec{\nabla} \hat{U}_m) \cdot ({}^t J_m^{-1} \vec{\nabla} \phi(\theta_m))] \, \sigma_m \, d\hat{x} d\hat{y}$$

$$= \sum_{m=1}^{M} \int_S f_{7,m} \phi(\theta_m) \sigma_m \, d\hat{x} d\hat{y},$$

${}^t J_m$ being the transpose of the Jacobian of θ_m.

Other equations involving the cross-points can be recovered from (4.2.1). To illustrate the situation at a 4d-type cross-point, we refer to Fig. 4.2.1, where V is the common vertex to the domains Ω_1, Ω_2, Ω_3 and Ω_4. The vectors $\vec{\nu}^{(m)}$, $1 \leq m \leq 4$, are perpendicular to the the common sides as the figure shows. From (4.2.1), assuming (4.2.2) and some regularity of the function U, one may also get

(4.2.14)
$$\frac{\partial U_1}{\partial \vec{\nu}^{(1)}} + \frac{\partial U_2}{\partial \vec{\nu}^{(2)}} + \frac{\partial U_3}{\partial \vec{\nu}^{(3)}} + \frac{\partial U_4}{\partial \vec{\nu}^{(4)}}$$

$$= \frac{\partial U_2}{\partial \vec{\nu}^{(1)}} + \frac{\partial U_4}{\partial \vec{\nu}^{(2)}} + \frac{\partial U_1}{\partial \vec{\nu}^{(3)}} + \frac{\partial U_3}{\partial \vec{\nu}^{(4)}} \qquad \text{at } V.$$

In this manner, at the cross-point we should satisfy the equation obtained by summing up four equations (like (4.2.3) evaluated at the point V), each one associated with the jump of the normal derivatives of U across Γ_1, Γ_2, Γ_3, Γ_4, respectively. Of course, the signs in (4.2.14) change depending on the way the vectors $\vec{\nu}^{(m)}$, $1 \leq m \leq 4$, have been oriented. As we did for (4.2.12), a similar relation can be written for (4.2.14) in terms of the partial derivatives of the functions $\hat{U}_m$, $1 \leq m \leq 4$, at the vertices of $\bar{S}$.

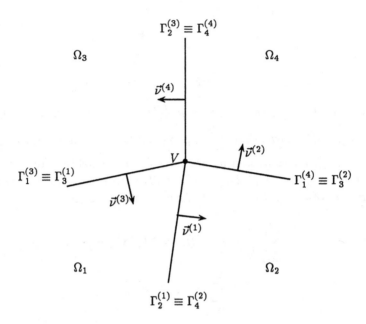

FIG. 4.2.1 - *Matching the subdomains at a cross-point.*

4.3   Approximation by spectral elements

We approximate the solution U, defined on $\bar{\Omega} = \cup_{m=1}^{M} \bar{\Omega}_m$, of the Poisson equation with homogeneous Dirichlet boundary conditions. More precisely, we approximate each function $\hat{U}_m$, $1 \leq m \leq M$, defined on the square $\bar{S}$, by a polynomial $\hat{q}_{n,m} \in \mathbf{P}_n^\star$ using the collocation method as described in the previous chapters. The degree $n \geq 2$ will be the same for all subdomains. The extension to the case of polynomials of different degrees in each subdomain will be analyzed later in section 4.5. The global approximation $Q_n : \bar{\Omega} \to \mathbf{R}$ of U is the patchwork of M functions $q_{n,m}$, $1 \leq m \leq M$, which are the restrictions of Q_n to $\bar{\Omega}_m$, such that $\hat{q}_{n,m} = q_{n,m}(\theta_m)$. We observe that the functions $q_{n,m}$, $1 \leq m \leq M$, may not be polynomials of $\mathbf{P}_n^\star$.

We start by noting that Q_n depends on $M(n+1)^2$ degrees of freedom. Moreover, a grid of $(n+1)^2$ points in $\bar{S}$ is tranformed into a grid on $\bar{\Omega}_m$, $1 \leq m \leq M$, via the mapping θ_m. It is easy to realize that any subdomain $\bar{\Omega}_m$ shares with a neighboring subdomain the same nodes at the interface. This is true for two reasons: first, that the number of nodes is the same for each subdomain; and second, that the functions θ_m, $1 \leq m \leq M$, are isoparametric

(see (4.1.1) and (4.1.2)). A node at a common side counts twice, and a 4d-type
cross-point counts four times. Taking for example the "house" of Fig. 4.1.1, in
Table 4.3.1 we give the number of distinct nodes (i.e. without counting their
multiplicity) for different choices of n.

n	*nodes inside* Ω	*boundary nodes*	*inteface nodes*	*total nodes in* $\bar{\Omega}$
3	120	84	48	204
4	232	112	70	344
5	380	140	92	520
6	564	168	124	732
7	784	196	136	980
8	1040	224	180	1264

TABLE 4.3.1 - *Number of nodes in the "house" with 18 subdomains for different n.*

We will use $M(n-1)^2$ nodes to collocate the differential equation inside
the subdomains. Hence, according to (4.2.4), for any $1 \le m \le M$, the
polynomial $\hat{q}_{n,m} \in \mathbf{P}_n^{\star}$ is required to satisfy the equation

$$(4.3.1) \qquad L_m \hat{q}_{n,m} = f_{7,m} \qquad \text{at the nodes inside } S.$$

The collocation grid in $\bar{S}$ depends on the coefficients $f_{k,m}$, $1 \le k \le 6$, of the
differential operator L_m and is determined as in section 2.2.

To ensure existence and uniqueness of Q_n, suitable conditions will be requested at the interfaces. First of all, Q_n must be continuous on $\bar{\Omega}$ and vanish
on $\partial\Omega$. Other conditions express the continuity of the normal derivatives (in
weak sense) across the interface sides and the cross-points. These relations
are the discrete counterparts of formulas (4.2.3) and (4.2.14). Enforcing the
boundary conditions on $\partial\Omega$ and the continuity of Q_n reduces the unknowns to
be determined. In fact, imposing the continuity of Q_n (which is equivalent to
assuming the continuity at the interface nodes) absorbs three degrees of freedom for a 4d-type cross-point, two degrees of freedom for a 3d-type cross-point,
and one degree of freedom for an interface point which is not a cross-point.
We are finally left with a number of conditions to be specified, which is equal

to the number of interface nodes considered with multiplicity 1. We take the inspiration from (4.2.12) to impose at those nodes the continuity of the derivatives in a weak sense. If $\Gamma \equiv \Gamma_2^{(r)} \equiv \Gamma_4^{(s)}$ is the interface between Ω_r and Ω_s, then the polynomials $\hat{q}_{n,r}$ and $\hat{q}_{n,s}$ both defined on $\bar{S}$ are required to satisfy the set of equations for $1 \leq i \leq n-1$:

$$(4.3.2) \qquad (\sigma_r \, L_r \hat{q}_{n,r})(1, \eta_i^{(n)}) \, \tilde{w}_n^{(n)} + (\sigma_s \, L_s \hat{q}_{n,s})(-1, \eta_i^{(n)}) \, \tilde{w}_0^{(n)}$$

$$+ \tfrac{1}{2}l(\Gamma)\frac{1}{\sigma_r} \left[\left(\nu_1 \frac{\partial \theta_{2,r}}{\partial \hat{y}} - \nu_2 \frac{\partial \theta_{1,r}}{\partial \hat{y}} \right) \frac{\partial \hat{q}_{n,r}}{\partial \hat{x}} \right.$$

$$\left. + \left(-\nu_1 \frac{\partial \theta_{2,r}}{\partial \hat{x}} + \nu_2 \frac{\partial \theta_{1,r}}{\partial \hat{x}} \right) \frac{\partial \hat{q}_{n,r}}{\partial \hat{y}} \right] (1, \eta_i^{(n)})$$

$$- \tfrac{1}{2}l(\Gamma)\frac{1}{\sigma_s} \left[\left(\nu_1 \frac{\partial \theta_{2,s}}{\partial \hat{y}} - \nu_2 \frac{\partial \theta_{1,s}}{\partial \hat{y}} \right) \frac{\partial \hat{q}_{n,s}}{\partial \hat{x}} \right.$$

$$\left. + \left(-\nu_1 \frac{\partial \theta_{2,s}}{\partial \hat{x}} + \nu_2 \frac{\partial \theta_{1,s}}{\partial \hat{x}} \right) \frac{\partial \hat{q}_{n,s}}{\partial \hat{y}} \right] (-1, \eta_i^{(n)})$$

$$= (\sigma_r f_{7,r})(1, \eta_i^{(n)}) \, \tilde{w}_n^{(n)} + (\sigma_s f_{7,s})(-1, \eta_i^{(n)}) \, \tilde{w}_0^{(n)},$$

where $\vec{\nu} \equiv (\nu_1, \nu_2)$, $\nu_1^2 + \nu_2^2 = 1$, is the normal vector to Γ pointing outside Ω_r and $l(\Gamma)$ is the length of Γ.

In the special case in which all the subdomains are squares having the sides of length 2 (thus θ_m is an *isometry* for $1 \leq m \leq M$), then (4.3.2) becomes for $1 \leq i \leq n-1$:

$$(4.3.3) \qquad - \Delta \hat{q}_{n,r}(1, \eta_i^{(n)}) \, \tilde{w}_n^{(n)} - \Delta \hat{q}_{n,s}(-1, \eta_i^{(n)}) \, \tilde{w}_0^{(n)}$$

$$+ \left(\frac{\partial \hat{q}_{n,r}}{\partial \hat{x}}(1, \eta_i^{(n)}) - \frac{\partial \hat{q}_{n,s}}{\partial \hat{x}}(-1, \eta_i^{(n)}) \right)$$

$$= f_{7,r}(1, \eta_i^{(n)}) \, \tilde{w}_n^{(n)} + f_{7,s}(-1, \eta_i^{(n)}) \, \tilde{w}_0^{(n)}.$$

This is a byproduct of the following variational problem (see also (3.2.1)):

$$(4.3.4) \qquad \sum_{m=1}^{M} \left[\sum_{i,j=0}^{n} [\vec{\nabla} \hat{q}_{n,m} \cdot \vec{\nabla} \phi(\theta_m)](\eta_j^{(n)}, \eta_i^{(n)}) \, \tilde{w}_i^{(n)} \tilde{w}_j^{(n)} \right]$$

$$= \sum_{m=1}^{M} \left[\sum_{i,j=0}^{n} [f_{7,m} \phi(\theta_m)](\eta_j^{(n)}, \eta_i^{(n)}) \, \tilde{w}_i^{(n)} \tilde{w}_j^{(n)} \right].$$

Note that (4.3.4) is recovered from (4.2.13) by setting $J_m = \begin{pmatrix} 1 & 0 \\ 0 & 1 \end{pmatrix}$, $\sigma_m = 1$, and replacing the integrals over $\bar{S}$ by quadratures. Note also the analogy between (4.3.3) and the imposition of Neumann boundary conditions in weak form expressed by equations (3.2.3)-(3.2.6). Another appropriate variational formulation leads to the more general equation (4.3.2).

The same idea is used to write the relations to be imposed at the cross-points. As we mentioned above, the equation relative to a $4d$-type cross-point is the result of a sum of four contributions, each one representing the continuity of the normal derivatives with respect to the neighboring sides. For the sake of simplicity, we only show the equation corresponding to a $4d$-type cross-point at the meeting of four 2×2 squares Ω_m, $1 \le m \le 4$. The subdomains are ordered as in Fig. 4.2.1 where the sides are pairwise orthogonal. Recalling that $\tilde{w}_0^{(n)} = \tilde{w}_n^{(n)} = 2/n(n+1)$, one has

$$
(4.3.5) \qquad \frac{2}{n(n+1)} \left[-\Delta \hat{q}_{n,1}(1,1) - \Delta \hat{q}_{n,2}(-1,1) \right.
$$

$$
\left. -\Delta \hat{q}_{n,3}(1,-1) - \Delta \hat{q}_{n,4}(-1,-1) \right]
$$

$$
+ \left(\frac{\partial \hat{q}_{n,1}}{\partial \hat{x}}(1,1) - \frac{\partial \hat{q}_{n,2}}{\partial \hat{x}}(-1,1) \right) + \left(\frac{\partial \hat{q}_{n,2}}{\partial \hat{y}}(-1,1) - \frac{\partial \hat{q}_{n,4}}{\partial \hat{y}}(-1,-1) \right)
$$

$$
+ \left(\frac{\partial \hat{q}_{n,3}}{\partial \hat{x}}(1,-1) - \frac{\partial \hat{q}_{n,4}}{\partial \hat{x}}(-1,-1) \right) + \left(\frac{\partial \hat{q}_{n,1}}{\partial \hat{y}}(1,1) - \frac{\partial \hat{q}_{n,3}}{\partial \hat{y}}(1,-1) \right)
$$

$$
= \frac{2}{n(n+1)} \left[f_{7,1}(1,1) + f_{7,2}(-1,1) + f_{7,3}(1,-1) + f_{7,4}(-1,-1) \right].
$$

The computation of the approximated solution Q_n amounts to the resolution of a linear system where the unknowns are the values of the polynomials $\hat{q}_{n,m}$, $1 \le m \le M$, at the collocation nodes, including the boundary and the interface nodes. The structure of the corresponding matrix is made up of M blocks, each one representing the set of equations in (4.3.1), coupled by the relations deriving from the interface conditions. We will not explicitly write such a matrix since we want to solve the linear system by an iterative procedure, which is described in the next section.

An interesting area of investigation is the analysis of convergence of Q_n to U for $n \to +\infty$. We do not present here any result in this direction. Some theory can be found in BERNARDI and MADAY (1989), and MADAY and RØNQUIST (1990).

4.4   An iterative algorithm for the domain decomposition method

We introduce an iterative procedure with the aim to find the solution of the linear system arising from the discretization of the Poisson equation by spectral elements. The purpose is to determine the values of Q_n at the interface nodes, which successively allow us to reconstruct the whole function Q_n, using the collocation problem satisfied by the the polynomials $\hat{q}_{n,m}$, $1 \leq m \leq M$. The algorithm can actually be interpreted as an iterative scheme for the so-called *capacitance* matrix with respect to the interface unknowns (see PROSKUROWSKI and WIDLUND (1976) and the references given therein, or DRYJA (1984)).

It is better to start with a simple example. We consider two subdomains ($M = 2$). Referring to Fig. 4.4.1, Ω_1 is a square of size 1×1, Ω_2 is a rectangle of size $\gamma \times 1$ (with $\gamma > 0$), and $\Gamma \equiv \Gamma_2^{(1)} \equiv \Gamma_4^{(2)}$ is their common side.

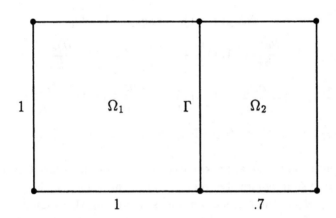

FIG. 4.4.1 - *Decomposition of a rectangle $\bar{\Omega}$ into two subdomains.*

Let $g_n : [-1, 1] \rightarrow \mathbf{R}$ be a polynomial of degree n in one variable such that $g_n(\pm 1) = 0$. Then, we construct a vector $\vec{g}_n$ of dimension $n - 1$ containing the values $g_n(\eta_i^{(n)})$, $1 \leq i \leq n-1$. Let $\hat{\chi}_{n,m} \in \mathbf{P}_n^*$, $1 \leq m \leq 2$, be two polynomials in $\bar{S}$ such that g_n is both the restriction of $\hat{\chi}_{n,1}$ to the side $\{1\} \times [-1, 1]$, and the restriction of $\hat{\chi}_{n,2}$ to the side $\{-1\} \times [-1, 1]$. We require that $\hat{\chi}_{n,1}$ and $\hat{\chi}_{n,2}$ vanish on the remaining part of ∂S. According to (4.3.1), inside S we impose

$$(4.4.1) \qquad (L_m \hat{\chi}_{n,m})(\eta_j^{(n)}, \eta_i^{(n)}) = f_{7,m}(\eta_j^{(n)}, \eta_i^{(n)}),$$

for $1 \leq i \leq n - 1$, $1 \leq j \leq n - 1$, where, recalling (4.2.5)-(4.2.11), one has

$$L_1 := -4\Delta, \quad L_2 := -4\left(\gamma^{-2}\frac{\partial^2}{\partial \hat{x}^2} + \frac{\partial^2}{\partial \hat{y}^2}\right), \quad \sigma_1 = \tfrac{1}{4} \text{ and } \sigma_2 = \tfrac{1}{4}\gamma.$$

At this point we define two other vectors $\vec{h}_{n,1}$ and $\vec{h}_{n,2}$ as follows

$$(4.4.2) \qquad \vec{h}_{n,1} := \left\{ [\sigma_1(L_1\hat{\chi}_{n,1} - f_{7,1})](1, \eta_i^{(n)}) \, \tilde{w}_n^{(n)} \right.$$

$$\left. + \frac{1}{4\sigma_1}\frac{\partial \hat{\chi}_{n,1}}{\partial \hat{x}}(1, \eta_i^{(n)}) \right\}_{1 \leq i \leq n-1}$$

$$(4.4.3) \qquad \vec{h}_{n,2} := \left\{ [\sigma_2(L_2\hat{\chi}_{n,2} - f_{7,2})](-1, \eta_i^{(n)}) \, \tilde{w}_0^{(n)} \right.$$

$$\left. - \frac{1}{4\sigma_2}\frac{\partial \hat{\chi}_{n,2}}{\partial \hat{x}}(-1, \eta_i^{(n)}) \right\}_{1 \leq i \leq n-1}.$$

Recalling (4.3.2) with $r = 1$ and $s = 2$ ($\vec{\nu} \equiv (1,0)$ and $l(\Gamma) = 1$), it is easy to realize that

$$(4.4.4) \qquad \vec{h}_n := \vec{h}_{n,1} + \vec{h}_{n,2} \equiv 0 \quad \Longleftrightarrow \quad \hat{\chi}_{n,m} = \hat{q}_{n,m}, \quad 1 \leq m \leq 2$$

In other words, the vector $\vec{h}_n$ vanishes if and only if the solutions of (4.4.1) are the polynomials corresponding to the approximation Q_n by spectral elements of the Poisson equation $-\Delta U = f$ in Ω, as described in section 4.3 (we remind that $Q_n(\theta_m) = \hat{q}_{n,m}$ in $\bar{S}$ for $1 \leq m \leq 2$). This characterization is also due to the uniqueness of Q_n.

We now determine the entries of $\vec{g}_n$ with the aim of obtaining $\vec{h}_n \equiv 0$. Note that the application $\vec{g}_n \to \vec{h}_n$ is *affine* and is linear if and only if $f = 0$. Its inverse is a discrete version of the Poincaré-Steklov operator (see section 3.5).

We define h_n to be the polynomial of degree n satisfying $h_n(\pm 1) = 0$ and assuming the values $\vec{h}_n$ at the nodes $\eta_i^{(n)}$, $1 \leq i \leq n - 1$. Moreover, we define the inner product

$$(4.4.5) \qquad < \vec{h}_n, \vec{g}_n > := \tfrac{1}{2}l(\Gamma)\sum_{i=0}^{n} h_n(\eta_i^{(n)})g_n(\eta_i^{(n)}) \, \tilde{w}_i^{(n)}$$

$$\approx \tfrac{1}{2}l(\Gamma)\int_{-1}^{1} h_n g_n \, dx,$$

where the weights are those of the Gauss-Lobatto formula given in $(A.2.3)$-$(A.2.4)$. We prove the following statement.

THEOREM 4.4.1 - *Let $f = 0$, then for any $n \geq 2$ one has*

(4.4.6) $$< \vec{h}_n, \vec{g}_n > \; \geq \; 0.$$

Moreover $< \vec{h}_n, \vec{g}_n > \; = \; 0$ if and only if $\vec{g}_n \equiv \vec{h}_n \equiv 0$.

Proof - We start by noting that $\hat{\chi}_{n,1}(1, \eta_i^{(n)}) = \hat{\chi}_{n,2}(-1, \eta_i^{(n)}) = g_n(\eta_i^{(n)})$, $1 \leq i \leq n - 1$. Then, due to (4.4.1), (4.4.2) and (4.4.3) with $f_{7,1} = f_{7,2} = 0$, we get:

(4.4.7) $$2[l(\Gamma)]^{-1} \; < \vec{h}_n, \vec{g}_n >$$

$$= \sigma_1 \sum_{i=0}^{n} [(L_1 \hat{\chi}_{n,1}) \hat{\chi}_{n,1}](1, \eta_i^{(n)}) \, \tilde{w}_i^{(n)} \tilde{w}_n^{(n)}$$

$$+ \frac{1}{4\sigma_1} \sum_{i=0}^{n} \left[\frac{\partial \hat{\chi}_{n,1}}{\partial \hat{x}} \hat{\chi}_{n,1} \right] (1, \eta_i^{(n)}) \, \tilde{w}_i^{(n)}$$

$$+ \sigma_2 \sum_{i=0}^{n} [(L_2 \hat{\chi}_{n,2}) \hat{\chi}_{n,2}](-1, \eta_i^{(n)}) \, \tilde{w}_i^{(n)} \tilde{w}_0^{(n)}$$

$$- \frac{1}{4\sigma_2} \sum_{i=0}^{n} \left[\frac{\partial \hat{\chi}_{n,2}}{\partial \hat{x}} \hat{\chi}_{n,2} \right] (-1, \eta_i^{(n)}) \, \tilde{w}_i^{(n)}$$

$$= \sigma_1 \sum_{j=0}^{n} \sum_{i=0}^{n} [(L_1 \hat{\chi}_{n,1}) \hat{\chi}_{n,1}](\eta_j^{(n)}, \eta_i^{(n)}) \, \tilde{w}_i^{(n)} \tilde{w}_j^{(n)}$$

$$+ \frac{1}{4\sigma_1} \sum_{i=0}^{n} \left[\frac{\partial \hat{\chi}_{n,1}}{\partial \hat{x}} \hat{\chi}_{n,1} \right] (1, \eta_i^{(n)}) \, \tilde{w}_i^{(n)}$$

$$+ \sigma_2 \sum_{j=0}^{n} \sum_{i=0}^{n} [(L_2 \hat{\chi}_{n,2}) \hat{\chi}_{n,2}](\eta_j^{(n)}, \eta_i^{(n)}) \, \tilde{w}_i^{(n)} \tilde{w}_j^{(n)}$$

$$- \frac{1}{4\sigma_2} \sum_{i=0}^{n} \left[\frac{\partial \hat{\chi}_{n,2}}{\partial \hat{x}} \hat{\chi}_{n,2} \right] (-1, \eta_i^{(n)}) \, \tilde{w}_i^{(n)}$$

$$= \sigma_1 \sum_{j=0}^{n} \sum_{i=0}^{n} \left[4 \left(\frac{\partial \hat{\chi}_{n,1}}{\partial \hat{x}} \right)^2 + 4 \left(\frac{\partial \hat{\chi}_{n,1}}{\partial \hat{y}} \right)^2 \right] (\eta_j^{(n)}, \eta_i^{(n)}) \, \tilde{w}_i^{(n)} \tilde{w}_j^{(n)}$$

$$+ \sigma_2 \sum_{j=0}^{n} \sum_{i=0}^{n} \left[4\gamma^{-2} \left(\frac{\partial \hat{\chi}_{n,2}}{\partial \hat{x}} \right)^2 + 4 \left(\frac{\partial \hat{\chi}_{n,2}}{\partial \hat{y}} \right)^2 \right] (\eta_j^{(n)}, \eta_i^{(n)}) \, \tilde{w}_i^{(n)} \tilde{w}_j^{(n)} \; \geq \; 0.$$

In particular, we used the quadrature formulas and the integration by parts as in the proof of theorem 1.3.1. We also recalled that $\hat{\chi}_{n,1}$ and $\hat{\chi}_{n,2}$ are vanishing on $\partial S - (\{1\}\times] - 1, 1[)$ and $\partial S - (\{-1\}\times] - 1, 1[)$, respectively. Note that the last term in (4.4.7) is an approximation to $\sum_{m=1}^{2} \int_{\Omega_m} |\vec{\nabla} U_m|^2 dx dy$.
Finally, if $< \vec{h}_n, \vec{g}_n >= 0$ then $\hat{\chi}_{n,1} = \hat{\chi}_{n,2} = 0$, which implies $\vec{h}_n \equiv \vec{g}_n \equiv 0$. This concludes the proof.

The theorem given above says that, assuming $f = 0$, the linear application $\vec{g}_n \to \vec{h}_n$ is positive definite. In addition, it is easy to show that the associated eigenvalues are real and strictly positive. Theorem 4.4.1 also provides a proof of the uniqueness of the spectral element solution Q_n for any given right-hand side function f.

We use (4.4.4) to build up an iterative procedure for the determination of Q_n. We are able to construct a sequence of vectors $\{\vec{g}_n^{(k)}\}_{k \in \mathbb{N}}$ which is expected to converge to the vector representing the values of Q_n at the interface Γ. At any step we have to solve $M = 2$ collocation problems. We start with an initial guess $\vec{g}_n^{(0)}$ and we construct the successive terms by the *Chebyshev iteration* method, which is well-suited for non-symmetric positive definite operators. Following MANTEUFFEL (1977), we first need to prescribe two parameters $c \geq 0$ and $d > 0$. Then, we compute

$$(4.4.8) \quad \begin{cases} \alpha_0 := d^{-1}, \quad \beta_0 := 0, \\[2ex] \vec{\delta}_n^{(k)} := -\alpha_k [\mathcal{R}_n^{-1} \vec{h}_n^{(k)}] + \beta_k \vec{\delta}_n^{(k-1)} \quad \text{for } k \in \mathbb{N}, \\[2ex] \vec{g}_n^{(k+1)} := \vec{g}_n^{(k)} + \vec{\delta}_n^{(k)} \quad \text{for } k \in \mathbb{N}, \\[2ex] \alpha_1 := 2d[2d^2 - c^2]^{-1} \text{ and } \alpha_{k+1} := \left[d - \left(\tfrac{1}{2}c\right)^2 \alpha_k\right]^{-1} \text{ for } k \geq 1, \\[2ex] \beta_{k+1} := d\alpha_{k+1} - 1 \quad \text{for } k \in \mathbb{N}, \end{cases}$$

where $\vec{h}_n^{(k)}$ is recovered as in (4.4.2) and (4.4.3), after finding the solutions $\hat{\chi}_{n,m}^{(k)}$, $1 \leq m \leq 2$, to the collocation problems having $\vec{g}_n^{(k)}$ as the vector of the Dirichlet data at the interface Γ. At the limit we have $\lim_{k \to +\infty} \vec{h}_n^{(k)} \equiv 0$ and $\lim_{k \to +\infty} \hat{\chi}_{n,m}^{(k)} = \hat{q}_{n,m}$, $1 \leq m \leq 2$.

In (4.4.8) the $(n-1)\times(n-1)$ matrix $\mathcal{R}_n$ is a preconditioner, introduced to speed up the convergence that otherwise would be very slow. The convergence

of the scheme is ensured when all the $n - 1$ eigenvalues associated with the
linear part of the application $\vec{g}_n \to \mathcal{R}_n^{-1} \vec{h}_n$ fall inside the ellipse in the complex
plane of foci $d - c$ and $d + c$, and passing through the origin. The semiaxes of
this ellipse are $a = d$ and $b = \sqrt{d^2 - c^2}$.

Taking into account the results of section 3.5, we choose $\mathcal{R}_n = \tilde{P}_n$, where
$\tilde{P}_n$ is the following tridiagonal matrix:

$$
(4.4.9) \quad \frac{5}{6}
\begin{bmatrix}
\frac{2}{\xi_2^{(n)} - \xi_1^{(n)}} & \frac{-1}{\eta_2^{(n)} - \xi_1^{(n)}} & & & & \\
\ddots & & \ddots & & & \\
& & \ddots & & \ddots & \\
& \frac{-1}{\xi_{i+1}^{(n)} - \eta_{i-1}^{(n)}} & \frac{2}{\xi_{i+1}^{(n)} - \xi_i^{(n)}} & \frac{-1}{\eta_{i+1}^{(n)} - \xi_i^{(n)}} & \\
& & \ddots & & \ddots & \\
& & & \ddots & & \ddots \\
& & & & \frac{-1}{\xi_n^{(n)} - \eta_{n-2}^{(n)}} & \frac{2}{\xi_n^{(n)} - \xi_{n-1}^{(n)}}
\end{bmatrix}
$$

Such a matrix is diagonal dominant and with positive eigenvalues. We
observe that the cost of evaluating $\tilde{P}_n^{-1}$ is negligible compared to that needed
for the implementation of the iterative algorithm (4.4.8).

The preconditioned eigenvalues are very well-behaved, as we can see by
running some experiments. We choose $\gamma = .7$, $f = 1$ and $\vec{g}_n^{(0)} \equiv 0$. We also
take $a = d = 2$ and $b = 1.5$, thus $c = \sqrt{1.75}$. We give in Table 4.4.1 the preconditioned residuals $\tilde{P}_n^{-1} \hat{h}_n^{(k)}$, evaluated in the norm $\sqrt{< \cdot, \cdot >}$ (see (4.4.5)),
for different values of n and k. These quantities decay with the same rate as
that of the norm of the error $\vec{g}_n^{(k+1)} - \vec{g}_n^{(k)}$ between two subsequent iterates.
We find in Fig. 4.4.2 the contour lines of the exact solution U. Although the
parameters c and d are not the ones providing the best convergence rate of the
scheme, the results of Table 4.4.1 show a good decay, which is almost indepen-
dent of n, denoting the effectiveness of the preconditioner $\mathcal{R}_n = \tilde{P}_n$. Moreover,
the convergence behavior does not substantially depend on the magnitude of
γ, provided there is no big difference between the area of Ω_1 and that of Ω_2
(say, for example, $.1 \leq \gamma \leq 5$).

k	$n = 4$	$n = 8$	$n = 12$	$n = 16$	$n = 20$
2	$.137E - 1$	$.243E - 1$	$.371E - 1$	$.384E - 1$	$.365E - 1$
3	$.441E - 1$	$.280E - 1$	$.674E - 2$	$.577E - 2$	$.120E - 1$
4	$.638E - 2$	$.103E - 1$	$.916E - 2$	$.270E - 2$	$.270E - 2$
5	$.545E - 2$	$.158E - 2$	$.325E - 2$	$.247E - 2$	$.155E - 3$
6	$.138E - 2$	$.181E - 2$	$.191E - 3$	$.111E - 2$	$.281E - 3$
7	$.570E - 2$	$.218E - 3$	$.379E - 3$	$.322E - 3$	$.219E - 3$
8	$.236E - 3$	$.203E - 3$	$.201E - 3$	$.368E - 4$	$.115E - 3$
9	$.467E - 4$	$.816E - 4$	$.385E - 4$	$.243E - 4$	$.493E - 4$
10	$.346E - 4$	$.991E - 5$	$.111E - 4$	$.199E - 4$	$.181E - 4$

TABLE 4.4.1 - *Preconditioned residuals using* $(4.4.8)$ *with* $c = \sqrt{1.75}$ *and* $d = 2$.

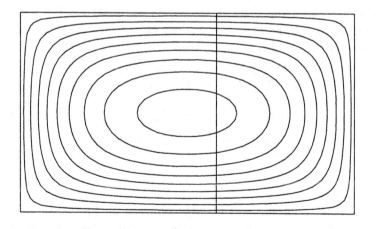

FIG. 4.4.2 - *Solution to* $-\Delta U = 1$ *on* Ω *decomposed into two subdomains with* $\gamma = .7$.

A similar iterative approach, although using a far more expensive preconditioner, was followed in CHAN and GOOVAERTS (1989), where the *conjugate
gradient* method was used in place of (4.4.8), providing an excellent rate of
covergence. The reason that we decided not to adopt the conjugate gradient
method here is that it requires the symmetric part of the operator $\vec{g}_n \rightarrow \tilde{P}_n^{-1}\vec{h}_n$
to be positive definite, a condition that does not seem to be realized in the
case of more subdomains with cross-points.

───── ◇ ─────

We are ready to study the decomposition of a set Ω in the general case.
For any $1 \leq m \leq M$, we denote by $g_{k,m}$, $1 \leq k \leq 4$, some prescribed n-
degree polynomials in one variable defined on the sides $[-1,1] \times \{-1\}$, $\{1\} \times
[-1,1]$, $[-1,1] \times \{1\}$, $\{-1\} \times [-1,1]$, respectively. These polynomials are re-
quired to satisfy the compatibility conditions: $g_{1,m}(1) = g_{2,m}(-1)$, $g_{2,m}(1) =
g_{3,m}(1)$, $g_{3,m}(-1) = g_{4,m}(1)$, $g_{4,m}(-1) = g_{1,m}(-1)$. Moreover, $g_{k,m}$ will be
identically zero when $\Gamma_k^{(m)}$ is a side belonging to $\partial\Omega$.
Recalling again (4.3.1), we impose

$$(4.4.10) \qquad \left(L_m \hat{\chi}_{n,m}\right)(\tau_{i,j}^{(n)}, v_{i,j}^{(n)}) = f_{7,m}(\tau_{i,j}^{(n)}, v_{i,j}^{(n)}),$$

for $1 \leq i \leq n-1$, $1 \leq j \leq n-1$.
In (4.4.10) the upwind grid nodes are chosen as, in section 2.2, depending on
the coefficients of the operator L_m. As done in (1.2.2), we provide the following
Dirichlet boundary conditions

$$(4.4.11) \qquad \begin{cases} \hat{\chi}_{n,m}(\eta_j^{(n)}, \eta_0^{(n)}) = g_{1,m}(\eta_j^{(n)}) & 0 \leq j \leq n-1, \\[2mm] \hat{\chi}_{n,m}(\eta_n^{(n)}, \eta_i^{(n)}) = g_{2,m}(\eta_i^{(n)}) & 0 \leq i \leq n-1, \\[2mm] \hat{\chi}_{n,m}(\eta_j^{(n)}, \eta_n^{(n)}) = g_{3,m}(\eta_j^{(n)}) & 1 \leq j \leq n, \\[2mm] \hat{\chi}_{n,m}(\eta_0^{(n)}, \eta_i^{(n)}) = g_{4,m}(\eta_i^{(n)}) & 1 \leq i \leq n, \end{cases}$$

so that the collocation problem (4.4.10) admits a unique solution.

Afterwards, we define other n-degree polynomials in one variable $h_{k,m}$:
$[-1,1] \rightarrow \mathbf{R}$ such that they are zero when $\Gamma_k^{(m)}$ is a side belonging to $\partial\Omega$,
and satisfy the compatibility conditions: $h_{1,m}(1) = h_{2,m}(-1)$, $h_{2,m}(1) =
h_{3,m}(1)$, $h_{3,m}(-1) = h_{4,m}(1)$, $h_{4,m}(-1) = h_{1,m}(-1)$. Moreover, when $\Gamma_k^{(m)}$ is
an interface side, we assume that:

(4.4.12) $\qquad h_{1,m}(\eta_j^{(n)}) := [\sigma_m(L_m\hat{\chi}_{n,m} - f_{7,m})](\eta_j^{(n)}, -1)\,\tilde{w}_0^{(n)}$

$$+ \tfrac{1}{2}l(\Gamma_1^{(m)})\left[\frac{1}{\sigma_m}\left(\nu_1^{(1)}\frac{\partial\theta_{2,m}}{\partial\hat{y}} - \nu_2^{(1)}\frac{\partial\theta_{1,m}}{\partial\hat{y}}\right)\frac{\partial\hat{\chi}_{n,m}}{\partial\hat{x}}\right](\eta_j^{(n)}, -1)$$

$$+ \tfrac{1}{2}l(\Gamma_1^{(m)})\left[\frac{1}{\sigma_m}\left(-\nu_1^{(1)}\frac{\partial\theta_{2,m}}{\partial\hat{x}} + \nu_2^{(1)}\frac{\partial\theta_{1,m}}{\partial\hat{x}}\right)\frac{\partial\hat{\chi}_{n,m}}{\partial\hat{y}}\right](\eta_j^{(n)}, -1)$$

for $1 \le j \le n - 1$,

(4.4.13) $\qquad h_{2,m}(\eta_i^{(n)}) := [\sigma_m(L_m\hat{\chi}_{n,m} - f_{7,m})](1, \eta_i^{(n)})\,\tilde{w}_n^{(n)}$

$$+ \tfrac{1}{2}l(\Gamma_2^{(m)})\left[\frac{1}{\sigma_m}\left(\nu_1^{(2)}\frac{\partial\theta_{2,m}}{\partial\hat{y}} - \nu_2^{(2)}\frac{\partial\theta_{1,m}}{\partial\hat{y}}\right)\frac{\partial\hat{\chi}_{n,m}}{\partial\hat{x}}\right](1, \eta_i^{(n)})$$

$$+ \tfrac{1}{2}l(\Gamma_2^{(m)})\left[\frac{1}{\sigma_m}\left(-\nu_1^{(2)}\frac{\partial\theta_{2,m}}{\partial\hat{x}} + \nu_2^{(2)}\frac{\partial\theta_{1,m}}{\partial\hat{x}}\right)\frac{\partial\hat{\chi}_{n,m}}{\partial\hat{y}}\right](1, \eta_i^{(n)})$$

for $1 \le i \le n - 1$,

(4.4.14) $\qquad h_{3,m}(\eta_j^{(n)}) := [\sigma_m(L_m\hat{\chi}_{n,m} - f_{7,m})](\eta_j^{(n)}, 1)\,\tilde{w}_n^{(n)}$

$$+ \tfrac{1}{2}l(\Gamma_3^{(m)})\left[\frac{1}{\sigma_m}\left(\nu_1^{(3)}\frac{\partial\theta_{2,m}}{\partial\hat{y}} - \nu_2^{(3)}\frac{\partial\theta_{1,m}}{\partial\hat{y}}\right)\frac{\partial\hat{\chi}_{n,m}}{\partial\hat{x}}\right](\eta_j^{(n)}, 1)$$

$$+ \tfrac{1}{2}l(\Gamma_3^{(m)})\left[\frac{1}{\sigma_m}\left(-\nu_1^{(3)}\frac{\partial\theta_{2,m}}{\partial\hat{x}} + \nu_2^{(3)}\frac{\partial\theta_{1,m}}{\partial\hat{x}}\right)\frac{\partial\hat{\chi}_{n,m}}{\partial\hat{y}}\right](\eta_j^{(n)}, 1)$$

for $1 \le j \le n - 1$,

(4.4.15) $\qquad h_{4,m}(\eta_i^{(n)}) := [\sigma_m(L_m\hat{\chi}_{n,m} - f_{7,m})](-1, \eta_i^{(n)})\,\tilde{w}_0^{(n)}$

$$+ \tfrac{1}{2}l(\Gamma_4^{(m)})\left[\frac{1}{\sigma_m}\left(\nu_1^{(4)}\frac{\partial\theta_{2,m}}{\partial\hat{y}} - \nu_2^{(4)}\frac{\partial\theta_{1,m}}{\partial\hat{y}}\right)\frac{\partial\hat{\chi}_{n,m}}{\partial\hat{x}}\right](-1, \eta_i^{(n)})$$

$$+ \tfrac{1}{2}l(\Gamma_4^{(m)})\left[\frac{1}{\sigma_m}\left(-\nu_1^{(4)}\frac{\partial\theta_{2,m}}{\partial\hat{x}} + \nu_2^{(4)}\frac{\partial\theta_{1,m}}{\partial\hat{x}}\right)\frac{\partial\hat{\chi}_{n,m}}{\partial\hat{y}}\right](-1, \eta_i^{(n)})$$

for $1 \le i \le n - 1$.

Here, $\vec{\nu}^{(k)} \equiv (\nu_1^{(k)}, \nu_2^{(k)})$, $[\nu_1^{(k)}]^2 + [\nu_2^{(k)}]^2 = 1$, $1 \le k \le 4$, denotes the normal vector to the side $\Gamma_k^{(m)}$ pointing outside Ω_m. For example, when Ω_m is a 2×2 square, $\vec{\nu}^{(1)} \equiv (0, -1)$, $f_{7,m} = 0$, and $g_{2,m} = g_{3,m} = g_{4,m} = 0$, then the quantities in (4.4.12) are exactly the entries of the vector given in (3.5.4).

At the corner point $(-1, -1)$ we require that:

$$(4.4.16) \quad h_{1,m}(-1) = h_{4,m}(-1) := [\sigma_m(L_m\hat{\chi}_{n,m} - f_{7,m})](-1, -1)\, \tilde{w}_0^{(n)}$$

$$+ \left[\frac{1}{\sigma_m}\left(\mu_1 \frac{\partial \theta_{2,m}}{\partial \hat{y}} - \mu_2 \frac{\partial \theta_{1,m}}{\partial \hat{y}} \right) \frac{\partial \hat{\chi}_{n,m}}{\partial \hat{x}} \right](-1, -1)$$

$$+ \left[\frac{1}{\sigma_m}\left(-\mu_1 \frac{\partial \theta_{2,m}}{\partial \hat{x}} + \mu_2 \frac{\partial \theta_{1,m}}{\partial \hat{x}} \right) \frac{\partial \hat{\chi}_{n,m}}{\partial \hat{y}} \right](-1, -1),$$

where $\mu_1 := \frac{1}{2}[l(\Gamma_1^{(m)})v_1^{(1)} + l(\Gamma_4^{(m)})v_1^{(4)}]$ and $\mu_2 := \frac{1}{2}[l(\Gamma_1^{(m)})v_2^{(1)} + l(\Gamma_4^{(m)})v_2^{(4)}]$. The reader can easily figure out the definitions corresponding to the other corners.

Now, we go back to the characterization given in (4.4.4), to impose the continuity of the weak normal derivatives across the interfaces. If, for instance, Ω_r and Ω_s are contiguous subdomains with interface $\Gamma \equiv \Gamma_2^{(r)} \equiv \Gamma_4^{(s)}$, thanks to (4.3.2) we would like to have

$$(4.4.17) \quad h_{2,r}(\eta_i^{(n)}) + h_{4,s}(\eta_i^{(n)}) = 0, \qquad 1 \le i \le n-1.$$

In the relation above, by accordingly adapting the signs, we recognize that the normal vector pointing outside Ω_r is opposite to the one pointing outside Ω_s. In the same way, at a $4d$-type cross-point, the counterpart of (4.4.17) will be expressed by the sum of four terms similar to (4.4.16).

By enforcing (4.4.17) and the equations corresponding to the other interfaces (including the cross-points), we should get $\hat{\chi}_{n,m} = \hat{q}_{n,m}$, $1 \le m \le M$. We define a vector $\vec{G}_n$ containing the values of the polynomials $\hat{\chi}_{n,m}$, $1 \le m \le M$, at the interface nodes, including the cross-points, counted with multiplicity one. Similarly, we define a vector $\vec{H}_n$, of the same dimension of $\vec{G}_n$, containing the difference of the weak normal derivatives (like the left-hand side of (4.4.17)) evaluated at each node belonging to the interfaces. Thus, there exists an operator, denoted by Υ_n, which allows us to pass from $\vec{G}_n$ to $\vec{H}_n$ by solving M collocation problems. The linear part of Υ_n is the capacitance matrix with respect to the interface unknowns of the linear system introduced in section 4.3. When the subdomains are all squares or rectangles, we can easily prove an extension of theorem 4.4.1, i.e. that Υ_n is positive definite. We can also conjecture that such a theorem is still valid for a general geometry. Here the difficulty in the proof relies on the fact that we cannot pass from the Gaussian quadratures to the exact integrals, since the coefficients $f_{k,m}$, $1 \le k \le 5$, in (4.2.5)-(4.2.9) are not constant, as in the case of the rectangles. Some hints for the theory may be taken from MADAY and RØNQUIST (1990).

As in (4.4.8), we are allowed to use the Chebyshev iteration method:

$$(4.4.18) \quad \begin{cases} \alpha_0 := d^{-1}, \quad \beta_0 := 0, \\[2ex] \vec{\Delta}_n^{(k)} := -\alpha_k[\mathcal{Q}_n^{-1}\vec{H}_n^{(k)}] + \beta_k\vec{\Delta}_n^{(k-1)} \quad \text{for } k \in \mathbf{N}, \\[2ex] \vec{G}_n^{(k+1)} := \vec{G}_n^{(k)} + \vec{\Delta}_n^{(k)} \quad \text{for } k \in \mathbf{N}, \\[2ex] \alpha_1 := 2d[2d^2 - c^2]^{-1} \text{ and } \alpha_{k+1} := \left[d - \left(\tfrac{1}{2}c\right)^2 \alpha_k\right]^{-1} \text{ for } k \geq 1, \\[2ex] \beta_{k+1} := d\alpha_{k+1} - 1 \quad \text{for } k \in \mathbf{N}, \end{cases}$$

where $\vec{H}_n^{(k)} = \Upsilon_n(\vec{G}_n^{(k)})$.

The vector $\vec{G}_n^{(0)}$ is given at the beginning of the process. We need to specify the preconditioning matrix $\mathcal{Q}_n$, which has an important role in the improvement of the convergence rate. For $k \in \mathbf{N}$, any interface side Γ is associated with a sub-vector $\vec{h}_n^{(k)}$ of length $n - 1$, obtained by extracting from the entries of $\vec{H}_n^{(k)}$ those corresponding to the nodes on Γ (endpoints excluded). Then, we apply the preconditioner on each interface side Γ by computing $\tilde{\mathcal{P}}_n^{-1}\vec{h}_n^{(k)}$, where $\tilde{\mathcal{P}}_n$ is the $(n - 1) \times (n - 1)$ matrix given in (4.4.9). At the entries of $\vec{H}_n^{(k)}$ corresponding to a $4d$-type cross-point V, the preconditioning procedure is obtained by dividing by $\frac{2}{3}[n(n+1)+1]$ (i.e. twice the value shown in (3.5.11)). In practice, after an appropriate ordering of the interface nodes, $\mathcal{Q}_n$ turns out to be a block diagonal matrix, where the $(n-1)\times(n-1)$ block $\tilde{\mathcal{P}}_n$ is repeated as many times as the number of interface sides, and the diagonal entries, in correspondence with the cross-points, are equal to $\frac{2}{3}[n(n + 1) + 1]$. Note that we will only need to store the matrix $\tilde{\mathcal{P}}_n$ and not the whole matrix $\mathcal{Q}_n$.

Here is a summary of the main steps of the iterative algorithm introduced above:

1) Organize a topological map of the domain $\bar{\Omega}$ as suggested in section 4.1, in order to know for any $1 \leq m \leq M$ the coordinates of Ω_m and the relationships between the subdomains of the partition;

2) Choose an initial guess $\vec{G}_n^{(k)}$, $k = 0$, which is used to impose the Dirichlet conditions at the interface nodes;

3) Construct the mapping θ_m in (4.1.1)-(4.1.2), and the coefficients of the operator L_m in (4.2.4), for any $1 \leq m \leq M$;

4) Find $\hat{\chi}_{n,m}^{(k)}$, $k \in \mathbb{N}$, for any $1 \leq m \leq M$, to be the polynomial solution to the collocation problem (4.4.10)-(4.4.11), where the Dirichlet boundary conditions on ∂S are partly related to the global boundary conditions on $\partial \Omega$ (if $\partial \Omega_m \cap \partial \Omega \neq \emptyset$), while at the nodes corresponding to the interfaces we take into account the vector $\vec{G}_n^{(k)}$;

5) Compute the weak normal derivatives of $\hat{\chi}_{n,m}^{(k)}$ on ∂S, for $1 \leq m \leq M$, and successively construct the vector $\vec{H}_n^{(k)} = \Upsilon_n(\vec{G}_n^{(k)})$ by reading the topological map to find out the connections between the subdomains;

6) Recover the preconditioned residuals $\mathcal{Q}_n^{-1}\vec{H}_n^{(k)}$ by applying the matrix $\tilde{\mathcal{P}}_n^{-1}$ to the $n-1$ entries of $\vec{H}_n^{(k)}$ corresponding to the nodes lined up along an interface side (excluding the endpoints), and dividing by $\frac{2}{3}[n(n+1)+1]$ the entries associated with the cross-points;

7) Evaluate the norm of $\mathcal{Q}_n^{-1}\vec{H}_n^{(k)}$ and, if less than a prescribed error, go to step *9*;

8) Update the vector $\vec{G}_n^{(k)}$ according to (4.4.18) and go to step *4*;

9) Stop the procedure. The functions $\hat{\chi}_{n,m}^{(k)}(\theta_m^{-1})$, $1 \leq m \leq M$, are the restrictions of Q_n to the set $\bar{\Omega}_m$, $1 \leq m \leq M$, where Q_n is the spectral element approximation of the Poisson equation in Ω.

The norm mentioned in step *7* may be the following one:

$$(4.4.19) \qquad \|\vec{G}_n\| := \left[\frac{1}{2} \sum_{m=1}^{M} \sum_{k=1}^{4} \left(\frac{1}{2} l(\Gamma_k^{(m)}) \sum_{i=1}^{n} [g_{k,m}(\eta_i^{(n)})]^2 \tilde{w}_i^{(n)} \right) \right]^{\frac{1}{2}},$$

which is a kind of L^2 norm at the interface, where the integrals are replaced by Gaussian quadratures. Formula (4.4.19) is a generalization of (4.4.5) to the case of more subdomains.

The preconditioned iterative scheme (4.4.18) has a very good convergence rate, although the number of iterations, needed to achieve a prescribed accuracy usually depends on the shape of the subdomains and the way they are assembled. In Table 4.4.2, for various n and k, we give the norm, evaluated using (4.4.19), of the preconditioned residuals in the case of the Poisson equation with $f = 1$. The domain Ω is the "house" decomposed in $M = 18$ subdomains (see Fig. 4.4.1). In these experiments we chose $c = \sqrt{1.75}$, $d = 2$ and $\vec{G}_n^{(0)} \equiv 0$. The convergence is not as good as in the case of two subdomains (see Table 4.4.1) due to the presence of the cross-points. Nevertheless, the decay of the error does not depend much on the degree n, pointing out the effectiveness of the preconditioning matrix. In order to further reduce the computational costs, several ameliorations will be proposed in section 4.5.

k	$n = 3$	$n = 4$	$n = 5$	$n = 6$
5	$.203E-1$	$.206E-1$	$.196E-1$	$.185E-1$
10	$.412E-2$	$.599E-2$	$.725E-2$	$.808E-2$
15	$.868E-3$	$.177E-2$	$.265E-2$	$.340E-2$
20	$.183E-3$	$.525E-3$	$.969E-3$	$.143E-2$
25	$.385E-4$	$.155E-3$	$.355E-3$	$.609E-3$
30	$.813E-5$	$.462E-4$	$.130E-3$	$.258E-3$
35	$.171E-5$	$.137E-4$	$.479E-4$	$.109E-3$

TABLE 4.4.2 - *Preconditioned residuals for the approximation of* $-\Delta U = 1$ *on the "house".*

Figs. 4.4.3 and 4.4.4 show the contour lines of the approximated solutions Q_n for $n = 3$ and $n = 6$, respectively. We recall that the continuity of the derivatives at the interfaces is imposed in a weak sense, which explains why Q_3 is not very smooth, being however a rather good global approximation. The number of unknowns involved in these computations are found in Table 4.3.1.

Finally, in Fig. 4.4.5, the reader finds the plot of the approximated solution obtained for $n = 6$ of the equation $-\Delta U = 1$ with homogeneous boundary conditions, defined on the domain of Fig. 4.1.4. At the $3d$-type cross-points the weak continuity is realized by summing up three terms similar to (4.4.16). The preconditioner acts exactly in the same manner described above.

Another preconditioned iterative procedure for the spectral approximation of the Poisson problem in complicated geometries has been investigated in NATARAJAN (1995). Some analogies can be found between our approach and those suggested in SMITH (1992) and CHAN, MATHEW and SHAO (1994).

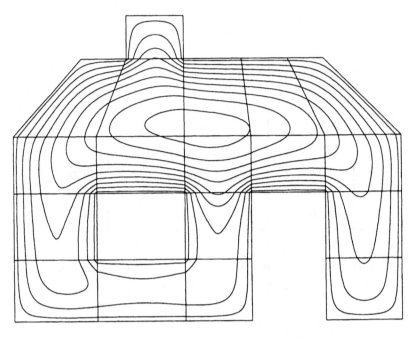

FIG. 4.4.3 - *Spectral element approximation to* $-\Delta U = 1$ *for* $n = 3$.

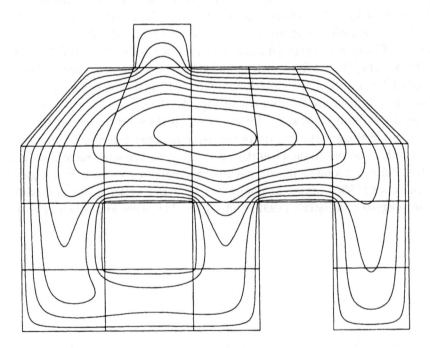

FIG. 4.4.4 - *Spectral element approximation to* $-\Delta U = 1$ *for* $n = 6$.

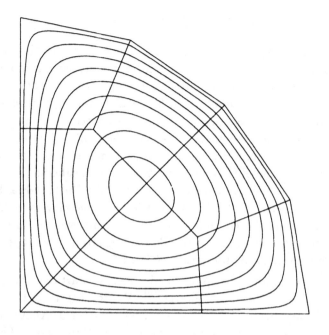

FIG. 4.4.5 - *Spectral element approximation to*
$-\Delta U = 1$ *on the "diamond" for* $n = 6$.

4.5 Improvements

At any step of the iterative algorithm introduced in the previous section, one must find the polynomials $\hat{\chi}_{n,m}$, $1 \leq m \leq M$, by solving the systems corresponding to M collocation problems like (4.4.10)-(4.4.11). The experiments of Table 4.4.1 and Table 4.4.2 have been carried-out by finding the solutions to these systems using the preconditioned Du Fort-Frankel method as described in sections 1.5 and 2.4, and letting the iterates converge up to the machine accuracy. Of course, we do not really need to compute these intermediate solutions with such an accuracy, while the Chebyshev iterative method (4.4.18) is still far from convergence. Thus, at any step $k \in \mathbf{N}$ of the Chebyshev iterative method, we can save a lot of computational time by stopping in advance the Du Fort-Frankel method, for instance, when the norm of the error between two successive iterates, evaluated using (1.4.5), is less than a prescribed quantity $\epsilon_n^{(k)} > 0$ decreasing with k. We suggest the following strategy: we set $\epsilon_n^{(0)}$

small enough and we take

$$(4.5.1) \qquad \epsilon_n^{(k+1)} := \delta \, \|\mathcal{Q}_n^{-1} \vec{H}_n^{(k)}\| \qquad \text{for } k \subset \mathbf{N},$$

where the norm $\| \cdot \|$ is given in (4.4.19) and $\delta = \frac{1}{10}$. In this way the collocation problems defined in each subdomain are solved incompletely. Such a choice, however, does not compromise the spectral accuracy of Q_n, since $\lim_{k \to +\infty} \epsilon_n^{(k)} = 0$. On the other hand, for small k, very few steps of the iterative solver used in each subdomain are sufficient to recover enough information at the interfaces to proceed with the scheme (4.4.18). This method is quite a time-saver. The formula proposed in (4.5.1) may be more or less effective depending on the shape of Ω and its decomposition in subdomains. We recommand checking that the number of iterations of the Du Fort-Frankel method actually increases with k. If this does not happen, we may loose the spectral accuracy (recall that the Du Fort-Frankel algorithm was initialized with the finite-differences discretization of the Poisson equation to be approximated). Such trouble is avoided by diminishing δ in (4.5.1).

Another way to reduce computational time is to start the Chebyshev iteration method with a better initial guess $\vec{G}_n^{(0)}$. We first construct the function Q_2 by running the algorithm (4.4.18) for $n = 2$ with $\vec{G}_2^{(0)} \equiv 0$. About 10 iterations are usually sufficient to get a stable solution up to three digits. This amounts to solve a so-called *coarse space* problem (see MANDEL (1994)). Then, we use the restriction of Q_2 to the interfaces (a kind of *wire basket*) to recover the entries of $\vec{G}_n^{(0)}$ (n is now supposed to be greater than 2). At any interface side we can use the interpolation formula

$$(4.5.2) \qquad r_n(\eta_i^{(n)}) = \sum_{j=0}^{2} r_2(\eta_j^{(2)}) \, \tilde{l}_j^{(2)}(\eta_i^{(n)})$$

$$= \tfrac{1}{2}\eta_i^{(n)}(\eta_i^{(n)} - 1)r_2(-1) + (1 - [\eta_i^{(n)}]^2)r_2(0) + \tfrac{1}{2}\eta_i^{(n)}(\eta_i^{(n)} + 1)r_2(1)$$

for $0 \leq i \leq n$, which holds true for any $r_n \in \mathbf{P}_n$, $n \geq 2$ (see also $(A.3.3)$). Of course, the preliminary part with $n = 2$ is very inexpensive and allows us to overcome the inertia of the scheme and avoid the use of the finer mesh when still unnecessary.

An improvement of the rate of convergence could be also obtained by trimmering the parameters c and d related to the size of the ellipse in the complex plane, which should contain the eigenvalues of the application $\vec{G}_n \to \mathcal{Q}_n^{-1} \vec{H}_n$. Despite the suggestions provided in MANTEUFFEL (1978), we were not able to devise any strategy for choosing the parameters in an optimal way. The difficulty lies in the way to control the minimum eigenvalue with an inexpensive procedure. However, the values proposed in section 4.4, i.e. $c = \sqrt{1.75}$

and $d = 2$, seem to ensure a satisfactory convergence rate independent of the geometry and the degree of the polynomials used in the approximation.

Other kinds of boundary conditions on $\partial\Omega$ can be also taken into account (see chapter three). In most of the cases, the imposition of these conditions involves each subdomain separately and has no influence on the structure of the algorithm (4.4.18), which only acts on the interface unknowns. We refer for an example to Fig. 4.5.1 ("the house with the smoking chimney"), corresponding to the approximated solution for $n = 6$ to the equation $-\Delta U = 1$, in the domain of Fig. 4.1.1. We assumed homogeneous Dirichlet boundary conditions on $\partial\Omega$, except for the horizontal side $\Gamma_3^{(14)}$ where homogeneous Neumann conditions were imposed. With the help of the mapping $\theta_{14} : \bar{S} \to \bar{\Omega}_{14}$ (see (4.1.1) and (4.1.2)) and recalling (3.2.5) and (3.4.2), one obtains the equations to be satisfied at the nodes of $\Gamma_3^{(14)}$, i.e.

$$(4.5.3) \quad (L_m \hat{\chi}_{n,m})(\eta_j^{(n)}, 1) + \frac{l(\Gamma_3^{(m)})}{2\tilde{w}_n^{(n)}\sigma_m^2} \frac{\partial\theta_{1,m}}{\partial\hat{x}} \frac{\partial\hat{\chi}_{n,m}}{\partial\hat{y}}(\eta_j^{(n)}, 1) = f_{7,m}(\eta_j^{(n)}, 1),$$

where $m = 14$ and $1 \leq j \leq n - 1$.

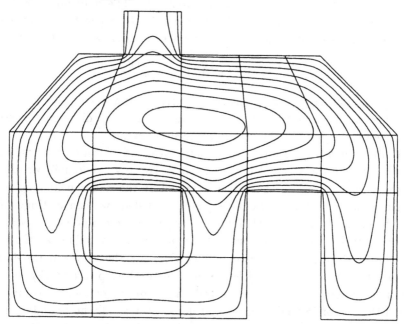

FIG. 4.5.1 - *Spectral element approximation for $n = 6$ to $-\Delta U = 1$ in Ω, with $U = 0$ on $\partial\Omega - \Gamma_3^{(14)}$ and $\frac{\partial}{\partial y}U = 0$ on $\Gamma_3^{(14)}$.*

The situation in which Neumann boundary conditions are imposed at the sides of two adjoining subdomains deserves further discussion. Let us take for an example the domains Ω_{11} and Ω_{16} and suppose to have homogeneous Neumann conditions on $\Gamma_3^{(11)}$ and $\Gamma_3^{(16)}$. We need to set up an equation for the vertex $V \equiv V_3^{(11)} \equiv V_4^{(16)} \in \partial\Omega$. It is preferable not to impose directly any condition on V, and consider in alternative V as a $4d$-type cross-point, assuming the existence of two other domains external to Ω. These fictitious domains give no contribution to the usual matching equations at V, since there we will sum up four weak derivatives instead of eight. In this way, using the algorithm (4.4.18), we only impose at V a Dirichlet-type condition which is updated step by step during the iteration process. At the "semi" cross-point V the preconditioning procedure is realized by dividing by $\frac{1}{3}[n(n+1)+1]$ (see also step 6 of the algorithm described in section 4.4). The same argument may apply for similar situations like the one of the node $V \equiv V_3^{(7)} \equiv V_2^{(9)} \equiv V_1^{(2)}$, when Neumann boundary conditions are requested on $\Gamma_2^{(7)}$ and $\Gamma_1^{(2)}$.

The multidomain approximation technique introduced in the previous section is very suitable for *parallel multiprocessors* machines. In fact, the computation of the solution in each subdomain can be assigned to a certain processor, and the exchange of information between the processors is only requested during the update of the values at the interfaces. The multiplication by the preconditioning matrix $\mathcal{Q}_n^{-1}$ can also be carried out in parallel. Considering, for simplicity, the case of the two domains, from (4.4.4) we clearly have

$$(4.5.4) \qquad \mathcal{Q}_n^{-1}\vec{h}_n \equiv \tilde{\mathcal{P}}_n^{-1}\vec{h}_{n,1} + \tilde{\mathcal{P}}_n^{-1}\vec{h}_{n,2}.$$

Thus, we can evaluate the two terms of the right-hand side of (4.5.4) independently and at the same time, since they are addressed to different processors. The same can be done in the case of more subdomains.

We describe a way to organize the numerical code, trying as much as possible to decouple the work of the processors. For any subdomain we allocate a memory space to store the values attained at the nodes belonging to the four sides plus the four vertices. Whenever needed, we can distinguish between boundary nodes on $\partial\Omega$, interface nodes and cross-points by virtue of the topological map (see section 4.1). In this way, the values at the nodes of the interface sides will be memorized twice, and the values at the $4d$-type cross-points will be memorized four times. At each step of (4.4.18), we solve in parallel the corresponding set of collocation problems and we compute in parallel the vector of the weak normal derivatives at all the boundary nodes of ∂S. Always in parallel, we apply the preconditioning matrix $\tilde{\mathcal{P}}_n^{-1}$ to the $n-1$ entries associated with each side, and we divide by $\frac{2}{3}[n(n+1)+1]$ the entries associated with the vertices. During this computation we do not really care to know if there are some normal derivatives that will be not used later. We are now ready for the only part that cannot be parallelized, which involves the exchange of information between the subdomains. In the way prescribed by the topological map, we have to take the difference between the preconditioned normal derivatives, which is used to correct by the same amount all the memory cells (that can vary from two to four) dedicated to a single interface node, in preparation for another step of the Chebyshev iteration method.

We finally give some hints for the generalization of the spectral element method to the case when different polynomial degrees are taken in each subdomain. We study the configuration of Fig. 4.4.1 and use polynomials of degree n_m in Ω_m, $1 \leq m \leq 2$, with $2 \leq n_1 \leq n_2$. An example of a grid in $\bar{\Omega}$ is shown in Fig. 4.5.2 ($n_1 = 3$, $n_2 = 4$). The spectral element approximation of U will be denoted by Q_{n_1,n_2}.

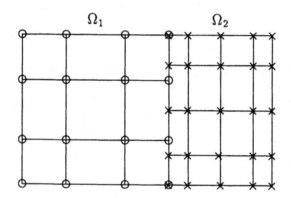

FIG. 4.5.2 - *Grid points in $\bar{\Omega}$ for $n_1 = 3$ and $n_2 = 4$.*

Let $g_{n_1} : [-1,1] \rightarrow \mathbf{R}$ be a polynomial of degree n_1 in one variable such that $g_{n_1}(\pm 1) = 0$. As in section 4.4, we construct a vector $\vec{g}_{n_1}$ of dimension $n_1 - 1$ containing the values $g_{n_1}(\eta_i^{(n_1)})$, $1 \leq i \leq n_1 - 1$. Then, we can build up two polynomials $\hat{\chi}_{n_m,m} \in \mathbf{P}_{n_m}^*$, $1 \leq m \leq 2$, defined in $\bar{S}$ such that g_{n_1} is both the restriction of $\hat{\chi}_{n_1,1}$ to the side $\{1\} \times [-1,1]$ and the restriction of $\hat{\chi}_{n_2,2}$ to the side $\{-1\} \times [-1,1]$. Moreover, for $1 \leq m \leq 2$, we require that

$$(4.5.5) \qquad (L_m \hat{\chi}_{n_m,m})(\eta_j^{(n_m)}, \eta_i^{(n_m)}) = f_{7,m}(\eta_j^{(n_m)}, \eta_i^{(n_m)}),$$

with $1 \leq i \leq n_m - 1$, $1 \leq j \leq n_m - 1$. We also impose that $\hat{\chi}_{n_m,m}$ vanishes on the sides of $\bar{S}$ corresponding to $\partial\Omega_m \cap \partial\Omega$. As in (4.4.2), we define

$$(4.5.6) \qquad \vec{h}_{n_1,1} := \left\{ [\sigma_1(L_1\hat{\chi}_{n_1,1} - f_{7,1})](1, \eta_i^{(n_1)}) \, \tilde{w}_{n_1}^{(n_1)} \right.$$

$$\left. + \frac{1}{4\sigma_1} \frac{\partial \hat{\chi}_{n_1,1}}{\partial \hat{x}} (1, \eta_i^{(n_1)}) \right\}_{1 \le i \le n_1 - 1}$$

In addition, we also define another vector of length $n_1 - 1$, i.e.

$$(4.5.7) \qquad \vec{h}_{n_1,2} := \left\{ \frac{1}{\tilde{w}_i^{(n_1)}} \sum_{j=1}^{n_2-1} \omega_j \, \tilde{l}_i^{(n_1)}(\eta_j^{(n_2)}) \, \tilde{w}_j^{(n_2)} \right\}_{1 \le i \le n_1 - 1}$$

where the Lagrange polynomials are given in $(A.3.1)$-$(A.3.2)$ and

$$(4.5.8) \quad \omega_j := [\sigma_2(L_2\hat{\chi}_{n_2,2} - f_{7,2})](-1, \eta_j^{(n_2)}) \, \tilde{w}_0^{(n_2)} - \frac{1}{4\sigma_2} \frac{\partial \hat{\chi}_{n_2,2}}{\partial \hat{x}}(-1, \eta_j^{(n_2)}),$$

for $1 \le j \le n_2 - 1$. Note that the vector in (4.5.7) is equal to the one in (4.4.3), when $n_1 = n_2$.

We point out that, for any $\phi \in \mathbf{P}_{n_1}$ with $\phi(\pm 1) = 0$, one has

$$(4.5.9) \qquad \sum_{i=1}^{n_1-1} \{\vec{h}_{n_1,2}\}_i \, \phi(\eta_i^{(n_1)}) \, \tilde{w}_i^{(n_1)} = \sum_{j=1}^{n_2-1} \omega_j \, \phi(\eta_j^{(n_2)}) \, \tilde{w}_j^{(n_2)}.$$

As in (4.4.4) the equation

$$(4.5.10) \qquad \vec{h}_{n_1} := \vec{h}_{n_1,1} + \vec{h}_{n_1,2} \equiv 0,$$

expresses the continuity of the weak normal derivatives at the interface nodes belonging to the grid in Ω_1 (the one related to the polynomial of lower degree). We observe that we required g_{n_1} to be a polynomial of degree n_1 in order to ensure the continuity along Γ of the function Q_{n_1,n_2}. Then, we note that a polynomial in $\mathbf{P}_{n_2}^\star$, which reduces to a polynomial of degree n_1 on $\{-1\} \times [-1,1]$, depends on $n_2(n_2 + 1) + n_1 + 1$ degrees of freedom. This explains why we only needed $n_1 - 1$ interface conditions, given by equation (4.5.10), to determine a unique function Q_{n_1,n_2}. The conditions to be imposed at the interface nodes are byproducts of the usual variational problem obtained by replacing the integrals by the quadratures in (4.2.1) (see FUNARO (1986)). The variational formulation leads to (4.5.10) by using (4.5.9) and the Lagrange basis $\phi = \tilde{l}_i^{(n_1)}$, $1 \le i \le n_1 - 1$, as a set of test functions. It is easy to adapt theorem 4.4.1 to the present situation, and also prove the convergence of Q_{n_1,n_2} to U for both n_1 and n_2 tending to $+\infty$.

From the variational formulation it is not difficult to derive the interface equations for the case of more subdomains. To avoid complicated notation, we prefer not to treat this extension.

By other techniques, such as the *mortar element* method, one is also able to handle *non-conforming* decompositions of the domain Ω, i.e. when a vertex of a subdomain may belong to the side of another subdomain. This approach, which does not require the continuity of the approximating function across the interfaces, allows for the matching of polynomials of different degrees and also the coupling of different discretization techniques (see BERNARDI, DEBIT and MADAY (1990)). The weak joint of the subdomains, obtained by a *least-squares* argument, makes the mortar element method particularly successful for non-conforming decompositions and domains with *moving parts*, i.e. decompositions which vary during the time evolution of the differential problem, as considered in ANAGNOSTOU, MADAY and PATERA (1997). The method has been implemented in the approximation of incompressible Navier-Stokes equations (see ACHDOU and PIRONNEAU (1996) and RØNQUIST (1996)) and stationary Maxwell equations. Some computational codes are available. We remark that the matching conditions of the mortar element method, relative to the conforming decomposition of Fig. 4.5.3, using a different number of degrees of freedom in each domain, coincide with those realized with the help of equation (4.5.10). As far as we know, *stabilization* techniques for transport-dominated equations, such as those we are discussing in this book, are not available for the mortar element method. For this reason, we forego a discussion of the method, although it is of relevance in the field of spectral elements, and refer the reader to MADAY, MAVRIPLIS and PATERA (1989), to the recent papers of BEN BELGACEM and MADAY (1994), BELHACHNI and BERNARDI (1994), and to the many references listed within these papers.

A general framework for domain decomposition problems, from which also the mortar element method can be derived, is the *three-field* method presented in BREZZI and MARINI (1992). Again, in the case of the conforming decompositions considered here, the spectral version of the three-field method leads to the discretizations examined above.

We finally provide a short list of standard references on domain decomposition methods for PDEs in 2-D or 3-D, emphasizing the interest for the preconditioned solvers. These are: BJØRSTAD and WINDLUND (1986), BRAMBLE, PASCIAK and SCHATZ (1986), SMITH (1992), XU (1992), DRYJA, SMITH and WINDLUND (1994), ACHDOU and KUZNETZOV (1995), PAVARINO and WINDLUND (1996). An updated bibliography is obtained by accessing *MGNet Bibliography* (http://casper.cs.yale.edu/mgnet/bib/mgnet.bib) by C.C. DOUGLAS and M.B. DOUGLAS.

4.6 Transport-diffusion equations in complex geometries

We start by studying the case when $\bar{\Omega}$ is the union of two subdomains, as in Fig. 4.4.1, and we approximate the solution to the equation (2.1.4), that is $-\epsilon\Delta U + \vec{\beta}\cdot\vec{\nabla}U = f_7$ in Ω, where $\epsilon > 0$ and $\vec{\beta} \equiv (\beta_1, \beta_2) \in \mathbf{R}^2$ are given, and $U = 0$ on $\partial\Omega$.

For $1 \leq m \leq 2$, as in (4.4.10), we replace (4.4.1) by

$$(4.6.1) \qquad (L_m\hat{\chi}_{n,m})(\tau_{i,j}^{(n)}, v_{i,j}^{(n)}) = f_{7,m}(\tau_{i,j}^{(n)}, v_{i,j}^{(n)}),$$

for $1 \leq i \leq n-1$, $1 \leq j \leq n-1$, where now we have

$$L_1 := -4\epsilon\Delta + 2\left(\beta_1\frac{\partial}{\partial\hat{x}} + \beta_2\frac{\partial}{\partial\hat{y}}\right) \qquad \text{and}$$

$$L_2 := -4\epsilon\left(\gamma^{-2}\frac{\partial^2}{\partial\hat{x}^2} + \frac{\partial^2}{\partial\hat{y}^2}\right) + 2\left(\gamma^{-1}\beta_1\frac{\partial}{\partial\hat{x}} + \beta_2\frac{\partial}{\partial\hat{y}}\right).$$

In (4.6.1) the collocation nodes depend on m and are determined according to the procedure described in section 2.2, using the coefficients of the operators L_1 and L_2, respectively. As mentioned in section 2.3, the collocation at the upwind grid is preferable for transport-dominated equations since it provides non oscillating approximated solutions and allows the implementation of very effective finite-differences preconditioning matrices. The operator L_1 is associated with the flux $\frac{1}{2}\epsilon^{-1}(\beta_1, \beta_2)$, a vector which expresses the direction and the intensity of the transport terms in S. In the same way, $\frac{1}{2}\epsilon^{-1}(\gamma\beta_1, \beta_2)$ is the flux corresponding to the operator L_2. We define the following quantities

$$(4.6.2) \qquad \Phi_N := \max\{\tfrac{1}{2}\epsilon^{-1}|\beta_1|, \tfrac{1}{2}\epsilon^{-1}\gamma|\beta_1|\}, \qquad \Phi_T := \tfrac{1}{2}\epsilon^{-1}\beta_2.$$

The first is a measure of the intensity of the component of the flux perpendicular to the sides $\{1\} \times [-1,1]$ and $\{-1\} \times [-1,1]$ of S, respectively, related to the interface Γ via the mappings θ_1 and θ_2 (see section 4.1). The second is the component of the flux tangential to those sides.

In order to use the iterative algorithm introduced in section 4.4, we define the two vectors $\vec{h}_{n,1}$ and $\vec{h}_{n,2}$:

$$(4.6.3) \qquad \vec{h}_{n,1} := \{[\sigma_1(L_1\hat{\chi}_{n,1} - f_{7,1})](1, \eta_i^{(n)})\,\tilde{w}_n^{(n)}$$

$$+ \frac{\epsilon}{4\sigma_1}\frac{\partial\hat{\chi}_{n,1}}{\partial\hat{x}}(1, \eta_i^{(n)})\}_{1\leq i\leq n-1}$$

(4.6.4) $\qquad \vec{h}_{n,2} := \{[\sigma_2(L_2\hat{\chi}_{n,2} - f_{7,2})](-1, \eta_i^{(n)})\, \tilde{w}_0^{(n)}$

$$- \frac{\epsilon}{4\sigma_2} \frac{\partial \hat{\chi}_{n,2}}{\partial \hat{x}}(-1, \eta_i^{(n)})\}_{1 \le i \le n-1}$$

which are very similar to those defined in (4.4.2) and (4.4.3). Therefore, the relation (4.4.4) still enforces the continuity of the weak normal derivatives, across the interface Γ, of the approximating spectral element solution.

Actually, the definitions (4.6.3) and (4.6.4) are associated with the discretization of equation (2.1.4) through its variational formulation: $\epsilon \int_\Omega \vec{\nabla} U \cdot \vec{\nabla}\phi\, dxdy + \int_\Omega \vec{\beta} \cdot \vec{\nabla} U \phi\, dxdy = \int_\Omega f\phi\, dxdy$. Note that the differential operator at the boundary nodes is automatically available when we use the iterative procedure described in section 2.4 (see step 9).

The Chebyshev iterative method (4.4.8) displays a good convergence behavior provided we adapt to the new situation the preconditioning matrix $\mathcal{R}_n$ of the capacitance matrix corresponding to the interface unknowns. Let us take for a moment $\vec{\beta} \equiv (\beta_1, 0)$, $\beta_1 \ne 0$, which is a vector perpendicular to Γ. Due to (4.6.2), we have $\Phi_T = 0$. Thus, we suggest taking

(4.6.5) $\qquad \mathcal{R}_n := \epsilon\left[(1 - e^{-n/\Phi_N})\, \tilde{\mathcal{P}}_n + (\Phi_N \alpha_n + \tfrac{3}{2}n\, e^{-n/\Phi_N})\, \mathcal{I}_n\right],$

where $\tilde{\mathcal{P}}_n$ is given in (4.4.9), $\mathcal{I}_n$ is the $(n-1) \times (n-1)$ identity matrix, and

(4.6.6) $\qquad \alpha_n := \dfrac{P_{n-1}(\xi_n^{(n)})}{(n+1)(1 - \xi_n^{(n)})} \approx .432,$

with $\xi_j^{(n)}$, $1 \le j \le n$, denoting the zeroes of the Legendre polynomial P_n. Note that for large Φ_N the diagonal of $\mathcal{R}_n$ becomes strongly dominant, while $\mathcal{R}_n \to \epsilon \tilde{\mathcal{P}}_n$ when Φ_N approaches zero.

Using the matrix given in (4.6.5) for the preconditioning of the Chebyshev iterative method, we get an excellent convergence behavior that does not seem to be very much affected by the parameters n and ϵ. The preconditioned residuals at the interface obtained by choosing for example $c = \sqrt{1.75}$ and $d = 2$ are comparable with those of Table 4.4.1. We run some tests in the case $f_7 = 1$, $\gamma = .7$, $\epsilon = .01$, $\vec{\beta} \equiv (1, 0)$. In Fig. 4.6.1 we show the corresponding approximated solutions for $n = 4$, $n = 8$ and $n = 16$. The latter case ensures a good resolution of the boundary layer. For smaller ϵ, by increasing the polynomial degree n, we get the same qualitative behavior. We observe that, when n is very small compared to Φ_N, the continuity of the derivatives at the interface is very weak (see the first plot of Fig. 4.6.1).

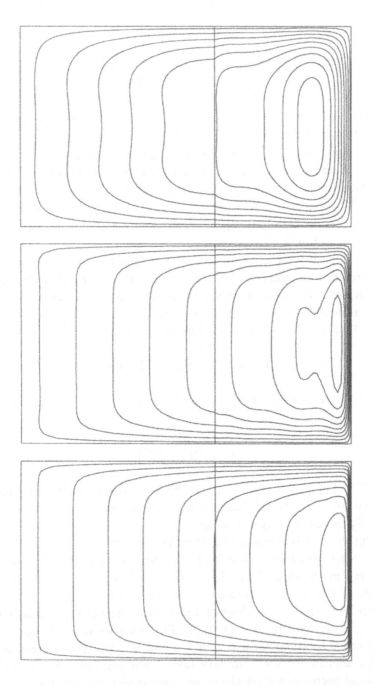

FIG. 4.6.1 - *Spectral element approximations for $n = 4$, $n = 8$ and $n = 16$ to the equation $-\frac{1}{100}\Delta U + \frac{\partial}{\partial x}U = 1$, with $U = 0$ on $\partial\Omega$.*

The reader finds in Fig. 4.6.2 the plot of the approximated solution obtained for $f_7 = 1$, $\gamma = .7$, $\epsilon = .0001$, $\vec{\beta} \equiv (1,0)$ and $n = 20$, where 12 iterations were sufficient to get an error at the interface of the order of 10^{-5}. Although the boundary layer is not yet perfectly resolved, the result of Fig. 4.6.2 seems to be more accurate than the one we would obtain in Ω with a single polynomial with double degree, i.e. $n = 40$. The numerical experiments are similar when $\vec{\beta} \equiv (-1,0)$, since Φ_N does not depend on the sign of the flux across Γ.

FIG. 4.6.2 - *Spectral element approximation for $n = 20$ to the equation* $-\frac{1}{10000}\Delta U + \frac{\partial}{\partial x}U = 1$, *with $U = 0$ on $\partial\Omega$.*

A different treatment is required in the case $\vec{\beta} \equiv (0,\beta_2)$, $\beta_2 \neq 0$, where now $\Phi_N = 0$. The difficulty in the case of a flux tangential to the interface, is not just a matter of finding a good preconditioner. When n is not too large to control the boundary layer, some oscillations developing along Γ may pollute the whole approximated solution on Ω. The situation is similar to the one observed by collocating the transport-diffusion equation in the single domain at the Legendre grid instead of using the upwind grid as described in chapter two (see FUNARO (1993)). Bad behavior is shown in Fig. 4.6.3 where the plots correspond to $f_7 = 1$, $\gamma = .7$, $\epsilon = .01$, $\vec{\beta} \equiv (0,1)$, with $n = 4$, $n = 6$ and $n = 8$, respectively.

To overcome this problem, we can also introduce some suitable upwind nodes on Γ depending on the magnitude of Φ_T. More precisely, we define in $]-1,1[$ the nodes $\zeta_i^{(n)}$, $1 \leq i \leq n - 1$, which are the zeroes of the equation

$$(4.6.7) \qquad P_n'(\zeta_i^{(n)}) - \Phi_T \, P_n(\zeta_i^{(n)}) = 0, \qquad 1 \leq i \leq n - 1,$$

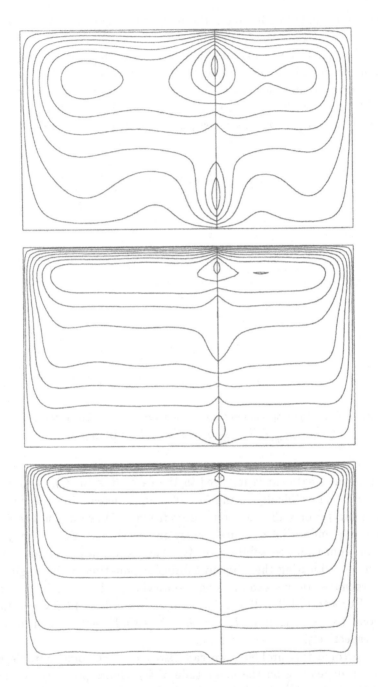

FIG. 4.6.3 - *Spectral element approximations for $n = 4$, $n = 6$ and $n = 8$ to the equation $-\frac{1}{100}\Delta U + \frac{\partial}{\partial y}U = 1$, with $U = 0$ on $\partial\Omega$.*

satisfying

$$(4.6.8) \quad \begin{cases} \xi_i^{(n)} < \zeta_i^{(n)} < \eta_i^{(n)} & \text{if } \Phi_T > 0, \\ \zeta_i^{(n)} = \eta_i^{(n)} & \text{if } \Phi_T = 0, \\ \eta_i^{(n)} < \zeta_i^{(n)} < \xi_{i+1}^{(n)} & \text{if } \Phi_T < 0. \end{cases}$$

Note the analogy with the relations (2.2.2), (2.2.3), and (2.2.7). Actually, (4.6.7) derives from (2.2.3) by taking $f_1 = f_4 = f_6 = 0$. In addition, we have $\zeta_i^{(n)} \to \xi_i^{(n)}$ if $\Phi_T \to +\infty$, and $\zeta_i^{(n)} \to \xi_{i+1}^{(n)}$ if $\Phi_T \to -\infty$. Successively, we map the nodes $(1, \zeta_i^{(n)}) \in \partial S$, $1 \le i \le n-1$, into Γ via the transformation $\theta_1 : \bar{S} \to \bar{\Omega}_1$ (see (4.1.1)-(4.1.2)). The same result is obtained by mapping the nodes $(-1, \zeta_i^{(n)}) \in \partial S$, $1 \le i \le n-1$, via the transformation $\theta_2 : \bar{S} \to \bar{\Omega}_2$.

In the new multidomain collocation scheme, we still satisfy the set of equations in (4.6.1), but the matching conditions at the interface are imposed by collocating at the modified interface grid. This yields

$$(4.6.9) \quad \vec{h}_n^* := \vec{h}_{n,1}^* + \vec{h}_{n,2}^* \equiv 0,$$

where we defined

$$(4.6.10) \quad \vec{h}_{n,1}^* := \left\{ [\sigma_1(L_1 \hat{\chi}_{n,1} - f_{7,1})](1, \zeta_i^{(n)}) \, \tilde{w}_n^{(n)} \right.$$

$$\left. + \frac{\epsilon}{4\sigma_1} \frac{\partial \hat{\chi}_{n,1}}{\partial \hat{x}}(1, \zeta_i^{(n)}) \right\}_{1 \le i \le n-1}$$

$$(4.6.11) \quad \vec{h}_{n,2}^* := \left\{ [\sigma_2(L_2 \hat{\chi}_{n,2} - f_{7,2})](-1, \zeta_i^{(n)}) \tilde{w}_0^{(n)} \right.$$

$$\left. - \frac{\epsilon}{4\sigma_2} \frac{\partial \hat{\chi}_{n,2}}{\partial \hat{x}}(-1, \zeta_i^{(n)}) \right\}_{1 \le i \le n-1}$$

We give the set of collocation nodes for the operator $-\frac{1}{20}\Delta + \frac{\partial}{\partial y}$ in the case $n = 4$ in Fig. 4.6.4.

We soon observe an improvement of the behavior of the approximated solutions. In Fig. 4.6.5, we show the results of the experiments obtained with the same data of Fig. 4.6.3, but this time we replaced (4.4.4) with (4.6.9).

Since one can prove that $\lim_{n \to +\infty} \zeta_i^{(n)} = \eta_i^{(n)}$ (see also (2.2.12)), for higher values of n (basically when the boundary layer starts to be well resolved) there is no sensible difference between the approximated solutions obtained by collocating the interface conditions at the Legendre nodes or at the upwind nodes.

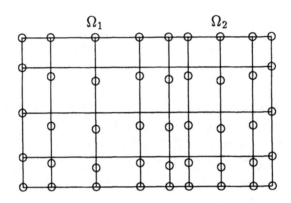

FIG. 4.6.4 - *Upwind grid for the multidomain collocation method corresponding to the operator* $-\frac{1}{20}\Delta + \frac{\partial}{\partial y}$ *and* $n = 4$.

To obtain the vectors in (4.6.10) and (4.6.11), we first compute the vectors $\vec{h}_{n,1}$ and $\vec{h}_{n,2}$ in (4.6.3) and (4.6.4). Then, we also need the quantities

$$[\sigma_1(L_1\hat{\chi}_{n,1} - f_{7,1})](1,-1)\,\tilde{w}_n^{(n)} \;+\; \frac{\epsilon}{4\sigma_1}\frac{\partial\hat{\chi}_{n,1}}{\partial\hat{x}}(1,-1),$$

$$[\sigma_1(L_1\hat{\chi}_{n,1} - f_{7,1})](1,1)\,\tilde{w}_n^{(n)} \;+\; \frac{\epsilon}{4\sigma_1}\frac{\partial\hat{\chi}_{n,1}}{\partial\hat{x}}(1,1),$$

$$[\sigma_2(L_2\hat{\chi}_{n,2} - f_{7,2})](-1,-1)\,\tilde{w}_0^{(n)} \;-\; \frac{\epsilon}{4\sigma_2}\frac{\partial\hat{\chi}_{n,2}}{\partial\hat{x}}(-1,-1),$$

$$[\sigma_2(L_2\hat{\chi}_{n,2} - f_{7,2})](-1,1)\,\tilde{w}_0^{(n)} \;-\; \frac{\epsilon}{4\sigma_2}\frac{\partial\hat{\chi}_{n,2}}{\partial\hat{x}}(-1,1),$$

and finally we can recover the entries of $\vec{h}^*_{n,1}$ and $\vec{h}^*_{n,2}$ by interpolation using $(A.3.3)$. This operation performed at the interface has a cost proportional to n^2.

When implementing the iterative method (4.4.8), we have to replace at any step the current vector $\vec{h}_n$ by the vector $\vec{h}^*_n$ (note that $\vec{h}_n \equiv \vec{h}^*_n$ when $\Phi_T = 0$). The problem of specifying the preconditioning matrix still remains. We set

$$(4.6.12) \qquad \mathcal{R}_n := \begin{cases} \epsilon\tilde{\mathcal{P}}_n \;+\; (\epsilon\Phi_T \max\{1, 1-\epsilon\})\,\mathcal{D}_n^+ & \text{if } \Phi_T > 0, \\[2ex] \epsilon\tilde{\mathcal{P}}_n \;+\; (\epsilon\Phi_T \max\{1, 1-\epsilon\})\,\mathcal{D}_n^- & \text{if } \Phi_T < 0, \end{cases}$$

where $\mathcal{D}_n^+$ and $\mathcal{D}_n^-$ are $(n-1) \times (n-1)$ diagonal matrices defined by:

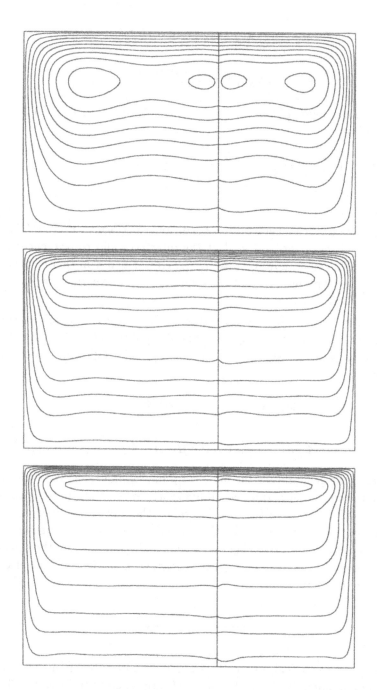

FIG. 4.6.5 - *Spectral element approximations using the upwind grid approach on* Γ
for the equation $-\frac{1}{100}\Delta U + \frac{\partial}{\partial y}U = 1$, $U = 0$ *on* $\partial\Omega$, *and* $n = 4$, $n = 6$, $n = 8$.

$$\mathcal{D}_n^+ := \frac{2}{n(n+1)} \text{ diag}\left\{ \frac{d}{dx}\tilde{l}_i^{(n)}(\xi_i^{(n)}) \right\}_{1 \le i \le n-1}$$

$$\mathcal{D}_n^- := \frac{2}{n(n+1)} \text{ diag}\left\{ \frac{d}{dx}\tilde{l}_i^{(n)}(\xi_{i+1}^{(n)}) \right\}_{1 \le i \le n-1}.$$

The Lagrange polynomials $\tilde{l}_i^{(n)}$, $1 \le i \le n-1$, are defined in $(A.3.1)$-$(A.3.2)$ and we have

$$(4.6.13) \qquad \left(\frac{d}{dx}\tilde{l}_i^{(n)}\right)(\xi) = \frac{(1-\xi^2)\, P_n'(\xi)}{n(n+1)\, P_n(\eta_i^{(n)})\, (\xi - \eta_i^{(n)})^2}, \qquad \xi \ne \eta_i^{(n)},$$

for $1 \le i \le n-1$.

Once again, such a choice of preconditioner performs very well (thanks also to the structure of the new grid at the interface) and convergence is rapidly achieved. This time, however, the parameters c and d in $(4.4.8)$ may depend a little on n. The best rate of convergence in our experiments was obtained with $c = \sqrt{2.75}$ and $d = 3$.

To handle the case of a general $\vec{\beta}$, we need only to combine the above procedures. The upwind interface nodes, derived from $(4.6.7)$, will still depend on the magnitude of Φ_T defined in $(4.6.2)$. The following preconditioning matrix

$$(4.6.14) \qquad \mathcal{R}_n := \epsilon\Big[(1 - e^{-n/\Phi_N})\, \tilde{\mathcal{P}}_n$$

$$+ (\Phi_N \alpha_n + \tfrac{3}{2}n\, e^{-n/\Phi_N})\mathcal{I}_n + (\Phi_T \max\{1, 1-\epsilon\})\mathcal{D}_n^{\pm}\Big],$$

extends to the general case the matrices proposed in $(4.6.5)$ and $(4.6.12)$. Concerning the convergence rate, we obtain excellent results, especially considering that for oblique transport terms, i.e. when the flux is not lined up with any one of the sides of rectangle Ω, the numerical problems are usually tougher, and less accurate solutions, together with a little slow down in the convergence, are commonly observed. The tricks suggested in section 4.5 can also help to reduce the computational time.

--- ◇ ---

We are able to deal now with the case of more subdomains Ω_m, $1 \le m \le M$. The polynomials $\hat{\chi}_{n,m}$, $1 \le m \le M$, are required to satisfy $(4.6.1)$. This time the coefficients $f_{4,m}$ and $f_{5,m}$ of the operator L_m in $(4.2.4)$ have to be modified as follows (see also $(4.2.8)$ and $(4.2.9)$):

$$(4.6.15) \qquad f_{4,m} := \frac{f_{2,m}}{\sigma_m} \left[\frac{\partial \theta_{1,m}}{\partial \hat{y}} \frac{\partial^2 \theta_{2,m}}{\partial \hat{x} \partial \hat{y}} - \frac{\partial \theta_{2,m}}{\partial \hat{y}} \frac{\partial^2 \theta_{1,m}}{\partial \hat{x} \partial \hat{y}} \right]$$

$$+ \frac{1}{\sigma_m} \left[\beta_1 \frac{\partial \theta_{2,m}}{\partial \hat{y}} - \beta_2 \frac{\partial \theta_{1,m}}{\partial \hat{y}} \right],$$

$$(4.6.16) \qquad f_{5,m} := \frac{f_{2,m}}{\sigma_m} \left[\frac{\partial \theta_{2,m}}{\partial \hat{x}} \frac{\partial^2 \theta_{1,m}}{\partial \hat{x} \partial \hat{y}} - \frac{\partial \theta_{1,m}}{\partial \hat{x}} \frac{\partial^2 \theta_{2,m}}{\partial \hat{x} \partial \hat{y}} \right]$$

$$+ \frac{1}{\sigma_m} \left[-\beta_1 \frac{\partial \theta_{2,m}}{\partial \hat{x}} + \beta_2 \frac{\partial \theta_{1,m}}{\partial \hat{x}} \right],$$

while $f_{1,m}$, $f_{2,m}$, and $f_{3,m}$ are multiplied by ϵ. At the interfaces, we impose the matching conditions, using the same approach adopted for the case of two subdomains. We examine the case when $\Gamma \equiv \Gamma_2^{(r)} \equiv \Gamma_4^{(s)}$ denotes the interface between the two contiguous domains Ω_r and Ω_s (similar arguments may be applied when $\Gamma \equiv \Gamma_3^{(r)} \equiv \Gamma_1^{(s)}$). Let $\vec{\nu} \equiv (\nu_1, \nu_2)$ be the unitary normal vector to Γ external to Ω_r. We define for $\hat{y} \in [-1, 1]$:

$$(4.6.17) \qquad \Phi_N^{(r,s)}(\hat{y}) :=$$

$$\max \left\{ \left| \frac{\nu_1}{\sigma_r f_{1,r}} \left[\beta_1 \frac{\partial \theta_{2,r}}{\partial \hat{y}} - \beta_2 \frac{\partial \theta_{1,r}}{\partial \hat{y}} \right] + \frac{\nu_2}{\sigma_r f_{3,r}} \left[-\beta_1 \frac{\partial \theta_{2,r}}{\partial \hat{x}} + \beta_2 \frac{\partial \theta_{1,r}}{\partial \hat{x}} \right] \right| (1, \hat{y}), \right.$$

$$\left. \left| \frac{\nu_1}{\sigma_s f_{1,s}} \left[\beta_1 \frac{\partial \theta_{2,s}}{\partial \hat{y}} - \beta_2 \frac{\partial \theta_{1,s}}{\partial \hat{y}} \right] + \frac{\nu_2}{\sigma_s f_{3,s}} \left[-\beta_1 \frac{\partial \theta_{2,s}}{\partial \hat{x}} + \beta_2 \frac{\partial \theta_{1,s}}{\partial \hat{x}} \right] \right| (-1, \hat{y}) \right\}$$

$$(4.6.18) \qquad \Phi_T^{(r,s)}(\hat{y}) :=$$

$$\frac{1}{2} \left\{ \left(\frac{\nu_2}{\sigma_r f_{1,r}} \left[\beta_1 \frac{\partial \theta_{2,r}}{\partial \hat{y}} - \beta_2 \frac{\partial \theta_{1,r}}{\partial \hat{y}} \right] - \frac{\nu_1}{\sigma_r f_{3,r}} \left[-\beta_1 \frac{\partial \theta_{2,r}}{\partial \hat{x}} + \beta_2 \frac{\partial \theta_{1,r}}{\partial \hat{x}} \right] \right) (1, \hat{y}) \right.$$

$$\left. + \left(\frac{\nu_2}{\sigma_s f_{1,s}} \left[\beta_1 \frac{\partial \theta_{2,s}}{\partial \hat{y}} - \beta_2 \frac{\partial \theta_{1,s}}{\partial \hat{y}} \right] - \frac{\nu_1}{\sigma_s f_{3,s}} \left[-\beta_1 \frac{\partial \theta_{2,s}}{\partial \hat{x}} + \beta_2 \frac{\partial \theta_{1,s}}{\partial \hat{x}} \right] \right) (-1, \hat{y}) \right\}.$$

These expressions generalize (4.6.2). According to (4.6.7), we construct the set of points

$$(4.6.19) \qquad P_n'(\zeta_i^{(n)}) - \Phi_T^{(r,s)}(\eta_i^{(n)}) P_n(\zeta_i^{(n)}) = 0, \qquad 1 \leq i \leq n - 1.$$

Then, in the case of the transport-diffusion equation, the relation (4.4.17) has the form

$$(4.6.20) \qquad h_{2,r}^*(\zeta_i^{(n)}) + h_{4,s}^*(\zeta_i^{(n)}) = 0, \qquad 1 \le i \le n-1,$$

where the polynomial $h_{2,r}^* \in \mathbf{P}_n$ takes the following values at the Legendre nodes

$$(4.6.21) \qquad h_{2,r}^*(\eta_i^{(n)}) := [\sigma_r(L_r \hat{\chi}_{n,r} - f_{7,r})](1, \eta_i^{(n)}) \, \tilde{w}_n^{(n)}$$

$$+ \tfrac{1}{2}\epsilon \, l(\Gamma) \left[\frac{1}{\sigma_r} \left(\nu_1 \frac{\partial \theta_{2,r}}{\partial \hat{y}} - \nu_2 \frac{\partial \theta_{1,r}}{\partial \hat{y}} \right) \frac{\partial \hat{\chi}_{n,r}}{\partial \hat{x}} \right] (1, \eta_i^{(n)})$$

$$+ \tfrac{1}{2}\epsilon \, l(\Gamma) \left[\frac{1}{\sigma_r} \left(-\nu_1 \frac{\partial \theta_{2,r}}{\partial \hat{x}} + \nu_2 \frac{\partial \theta_{1,r}}{\partial \hat{x}} \right) \frac{\partial \hat{\chi}_{n,r}}{\partial \hat{y}} \right] (1, \eta_i^{(n)}), \qquad 0 \le i \le n.$$

Note that $h_{2,r}^*$, obtained for $\epsilon = 1$ and $\vec{\beta} \equiv 0$, does not coincide with $h_{2,m}$, $m = r$, given in (4.4.13), since these two polynomials attain different values at the endpoints of the interval $[-1, 1]$. The definition of $h_{4,s}^*$ is very similar. The same treatment is reserved for the other interface sides and the equations to be satisfied at the cross-points are imposed in the usual manner.

For the numerical implementation, we use (4.4.18). The method is preconditioned by applying the matrix $\mathcal{R}_n^{-1}$ to the $n - 1$ unknowns corresponding to a certain interface side, where Φ_N and Φ_T are now vectors depending on the magnitude of the flux across that side. For instance, when $\Gamma \equiv \Gamma_2^{(r)} \equiv \Gamma_4^{(s)}$, we take $\{\Phi_N^{(r,s)}(\eta_i^{(n)})\}_{1 \le i \le n-1}$ and $\{\Phi_T^{(r,s)}(\eta_i^{(n)})\}_{1 \le i \le n-1}$, obtained from (4.6.17)-(4.6.18), to replace Φ_N and Φ_T in (4.6.14), respectively. At any cross-point we divide by the factor $\tfrac{2}{3}\epsilon[n(n+1)+1] + \epsilon\Phi$, where $\Phi \ge 0$ gives a measure of the magnitude of the flux. For instance, referring to Fig. 4.2.1 and recalling (4.6.17), we set

$$\Phi := \max\left\{ \sqrt{[\Phi_N^{(1,3)}(1)]^2 + [\Phi_N^{(1,2)}(1)]^2}, \; \sqrt{[\Phi_N^{(1,2)}(1)]^2 + [\Phi_N^{(2,4)}(-1)]}, \right.$$

$$\left. \sqrt{[\Phi_N^{(2,4)}(-1)]^2 + [\Phi_N^{(3,4)}(-1)]^2}, \; \sqrt{[\Phi_N^{(3,4)}(-1)]^2 + [\Phi_N^{(1,3)}(1)]^2} \right\}.$$

Hence, the procedure to be followed is equal to the one outlined in section 4.4, except that between steps 5 and 6 we have to interpolate the weak normal derivatives at the boundary of S from the Legendre nodes to the upwind nodes, determined at each interface side according to the size of the transverse flux.

We are finally able to show the results of some numerical tests in Figs. 4.6.6 and 4.6.7, where a western wind ($\vec{\beta} \equiv (0, 1)$, $\epsilon = \frac{1}{100}$) and a south-eastern wind ($\vec{\beta} \equiv (-1, -1)$, $\epsilon = \frac{1}{100}$) are respectively blowing on the "house".

Some other experiments, obtained without using the upwind grid approach, are reported in PASQUARELLI and QUARTERONI (1994).

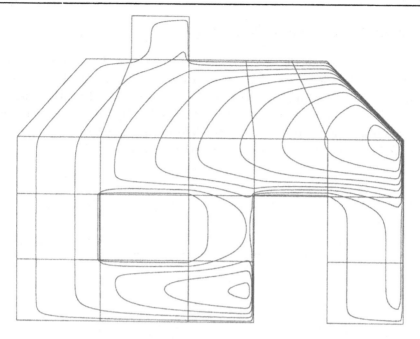

FIG. 4.6.6 - *Spectral element approximation to* $-\frac{1}{100}\Delta U + \frac{\partial}{\partial x}U = 1$,
with $U = 0$ *on* $\partial\Omega$, *for* $n = 16$.

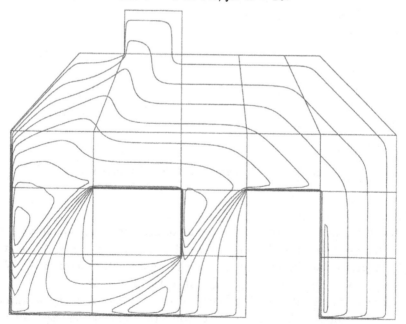

FIG. 4.6.7 - *Spectral element approximation to* $-\frac{1}{100}\Delta U - \frac{\partial}{\partial x}U - \frac{\partial}{\partial y}U = 1$,
with $U = 0$ *on* $\partial\Omega$, *for* $n = 16$.

4.7 A model problem in Electrostatics

We are going to apply the spectral element method to find the potential U, relative to an electric field $\vec{E}$, in a square box B of dimension 3×3 of a dielectric medium. Inside B, a smaller square A of dimension 1×1 of another dielectric material is contained. As we can see in Fig. 4.7.1, the potential is known at two sides of the boundary of B and $\delta > 0$ is a constant. The lower and upper sides are insulated. Both the dielectrics are isotropic.

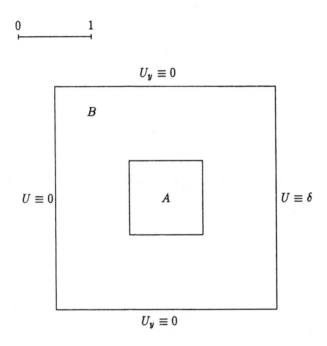

FIG. 4.7.1 - *Displacement of the two dielectrics.*

Due to symmetry, we consider only half of the computational domain which will be further decomposed in $M = 6$ subdomains Ω_k, $1 \le k \le M$, as in Fig. 4.7.2. In particular, Ω_3 corresponds to the upper half of square A. We set: $\bar{\Omega} := \cup_{1 \le k \le 6} \bar{\Omega}_k$, $\bar{\Omega}_B := \bar{\Omega}_1 \cup \bar{\Omega}_2 \cup \bar{\Omega}_4 \cup \bar{\Omega}_5 \cup \bar{\Omega}_6$, $\Omega_A := \Omega_3$.

Taking into account that $\vec{E} \equiv -\vec{\nabla}U$, where $U : \bar{\Omega} \to \mathbf{R}$ is the potential, we have $-\Delta U = \mathrm{div}\,\vec{E}$, which brings us to solve the following Poisson equations:

(4.7.1) $- C_B\,\Delta U_k\ =\ 1 \quad$ in Ω_k for $k = 1, 2, 4, 5, 6,$

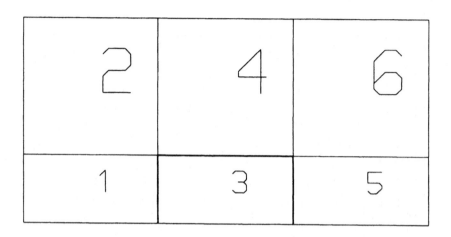

FIG. 4.7.2 - *Decomposition of the domain Ω.*

(4.7.2) $$- C_A \, \Delta U_3 \; = \; 1 \quad \text{in } \Omega_3.$$

In (4.7.1) and (4.7.2), U_k denotes the restriction of U to $\bar{\Omega}_k$, $1 \leq k \leq 6$. The positive constants C_B and C_A are proportional to the *relative permittivity* of the two dielectrics. For the derivation and physical interpretation of the above equations, numerous textbooks on electromagnetism exist. We mention in particular that of BLEANEY and BLEANEY (1976).

Using the notation introduced in section 4.1 to indicate the sides of the subdomains, the boundary conditions are as follows

(4.7.3) $\quad U = 0 \quad$ on $\Gamma_4^{(1)}$, $\Gamma_4^{(2)}$, $\quad$ and $\quad U = \delta \quad$ on $\Gamma_2^{(5)}$, $\Gamma_2^{(6)}$,

(4.7.4) $\quad \dfrac{\partial U}{\partial y} = 0 \quad$ on $\Gamma_3^{(2)}$, $\Gamma_3^{(4)}$, $\Gamma_3^{(6)}$, $\Gamma_1^{(1)}$, $\Gamma_1^{(3)}$, $\Gamma_1^{(5)}$.

Finally, we need to specify the interface conditions between the two dielectric materials, i.e.

(4.7.5) $\quad C_B \dfrac{\partial U_1}{\partial x} = C_A \dfrac{\partial U_3}{\partial x} \quad$ on $\Gamma_2^{(1)} \equiv \Gamma_4^{(3)}$,

(4.7.6) $\quad C_B \dfrac{\partial U_5}{\partial x} = C_A \dfrac{\partial U_3}{\partial x} \quad$ on $\Gamma_4^{(5)} \equiv \Gamma_2^{(3)}$,

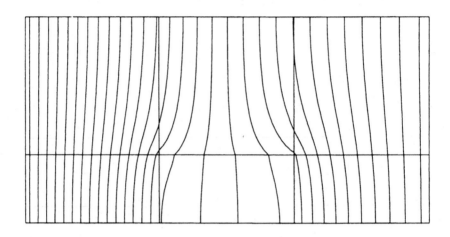

FIG. 4.7.3 - *Approximated electric potential for $C_B = 5 > C_A = 1$ and $n = 6$.*

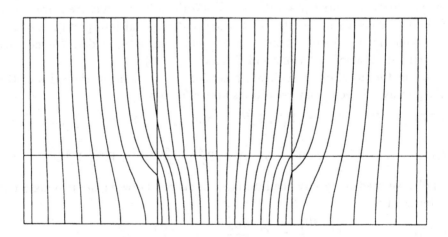

FIG. 4.7.4 - *Approximated electric potential for $C_B = 1 < C_A = 5$ and $n = 6$.*

(4.7.7) $$C_B \frac{\partial U_4}{\partial y} = C_A \frac{\partial U_3}{\partial y} \quad \text{on} \quad \Gamma_1^{(4)} \equiv \Gamma_3^{(3)}.$$

In addition, we shall require that U is a C^1 function on the domain $\bar{\Omega}_B$, an assumption which leads to the usual interface conditions studied in the previous sections. Note however that the global solution U is only C^0 in $\bar{\Omega}$.

At this stage, the construction of the approximation scheme and its numerical solution do not present problems. To treat the Neumann-type boundary conditions we can follow the instructions provided in chapter three (see also (4.5.3)). At the corner points of $\bar{\Omega}$ we apply the Dirichlet conditions in (4.7.3). Dealing with the equations to be imposed at the intersection between the dielectrics is a little more delicate. They can actually be derived from a variational formulation associated with the equations (4.7.1) and (4.7.2), and are similar to the usual interface conditions except that the normal derivatives are weighted in a different way depending on whether they are evaluated in Ω_B or Ω_A. For example, we analyze the case of the side $\Gamma_2^{(1)} \equiv \Gamma_4^{(3)}$, where (4.4.17) has to be replaced by

(4.7.8) $$C_B\, h_{2,1}(\eta_i^{(n)}) + C_A\, h_{4,3}(\eta_i^{(n)}) = 0, \qquad 1 \le i \le n-1.$$

In the iterative method (4.4.18), we use as preconditioner the matrix $C\tilde{\mathcal{P}}_n$, where $C := C_B$ for the interface sides contained in Ω_B, and $C := \max\{C_B, C_A\}$ for the interface sides belonging to $\bar{\Omega}_B \cap \bar{\Omega}_A$.

The results of some numerical tests are available. We take $\delta = 10$. In Fig. 4.7.3, we find the contour lines of the approximated potential obtained for $C_B = 5$, $C_A = 1$ and using polynomials of degree $n = 6$. Another solution, relative to the case $C_B = 1$ and $C_A = 5$ is shown in Fig. 4.7.4. Both the solutions are increasing functions of the variable x. These experiments are in agreement with the physics model.

Time
Discretization

We turn our attention to the approximation of partial differential equations where the solutions also depend on time. We discuss some well-known boundary-value problems and consider their discretization both in space and time.

5.1 Time-dependent advection-diffusion problems

In the time interval $[0, T]$, $T > 0$, we consider the evolution of the function U on the set $\bar{\Omega}$, according to the differential model

$$(5.1.1) \qquad \frac{\partial U}{\partial t} - \epsilon \Delta U + \vec{\beta} \cdot \vec{\nabla} U = 0 \qquad \text{in } \Omega \times]0, T],$$

where the constant $\epsilon > 0$ and the vector $\vec{\beta} \equiv (\beta_1, \beta_2) : \Omega \to \mathbf{R}^2$, are given. The solution $U(x, y, t)$, $(x, y) \in \bar{\Omega}$, $t \in [0, T]$, is required to satisfy the following initial and boundary conditions

$$(5.1.2) \qquad U(x, y, 0) = U_0(x, y), \qquad \forall (x, y) \in \Omega,$$

$$(5.1.3) \qquad U(x, y, t) = 0, \qquad \forall (x, y) \in \partial\Omega, \quad \forall t \in [0, T],$$

U_0 being a given function in Ω.

Assuming some regularity for the data, one can prove existence and uniqueness of U (see DAUTRAY and LIONS (1988-1993), XVIII, section 5.1). In addition, we have the following stability result:

THEOREM 5.1.1 - *Let* $C := \sup_\Omega |\text{div}\vec{\beta}|$, *then we have*

(5.1.4) $$\int_\Omega U^2(x,y,t)\,dxdy \leq e^{Ct} \int_\Omega U_0^2(x,y)\,dxdy, \qquad \forall t \in [0,T].$$

Moreover, if $\text{div}\vec{\beta} \leq 0$ *in* Ω, *we get*

(5.1.5) $$\frac{d}{dt}\int_\Omega U^2(x,y,t)\,dxdy \leq 0, \qquad \forall t \in [0,T].$$

The theorem above is easily proven by writing (5.1.1)-(5.1.3) in variational form, i.e.

(5.1.6) $$\int_\Omega \frac{\partial U}{\partial t}\phi\,dxdy + B(U,\phi) = 0,$$

with

$$B(U,\phi) := \epsilon \int_\Omega \vec{\nabla}U \cdot \vec{\nabla}\phi\,dxdy + \int_\Omega \vec{\beta}\cdot\vec{\nabla}U\ \phi\,dxdy,$$

where ϕ is a test function.

By taking $\phi = U$ and considering that $\int_\Omega [\frac{\partial}{\partial t}U]U\,dxdy = \frac{1}{2}\frac{d}{dt}\int_\Omega U^2 dxdy$, we get (5.1.5) since $B(U,U) \geq \int_\Omega (\vec{\beta}\cdot\vec{\nabla}U)U\,dxdy = -\frac{1}{2}\int_\Omega U^2\text{div}\vec{\beta}dxdy \geq 0$, if $\text{div}\vec{\beta} \leq 0$. The inequality (5.1.4) is a consequence of the *Gronwall lemma*.

Concerning the numerical approximation, we begin by discretizing with respect to the variable t. We denote by $\delta_t > 0$ the time step and by $U^{(k)}(x,y)$, $(x,y) \in \bar{\Omega}$ the approximated solution at the time $t_k := k\delta_t$, with $k \in \mathbf{N}$ such that $t_k \leq T$. Using the *trapezoidal rule*, we obtain

(5.1.7) $$\delta_t^{-1}\Big[U^{(k)} - U^{(k-1)}\Big] + \frac{1}{2}\Big[-\epsilon\Delta U^{(k)} + \vec{\beta}\cdot\vec{\nabla}U^{(k)}\Big]$$

$$+ \frac{1}{2}\Big[-\epsilon\Delta U^{(k-1)} + \vec{\beta}\cdot\vec{\nabla}U^{(k-1)}\Big] = 0 \qquad k \geq 1,$$

with

(5.1.8) $$U^{(0)} = U_0 \quad \text{in } \Omega, \qquad U^{(k)} = 0 \quad \text{on } \partial\Omega \text{ for } k \geq 0.$$

It turns out that, if $\text{div}\vec{\beta} \leq 0$, the method (5.1.7) is second-order accurate, i.e. the error between the exact and the approximated solutions at the time T decays as δ_t^2. Furthermore, no upper bound to the size of the time step is required. Actually, proof of the *unconditional stability* of the method is soon obtained by multiplying (5.1.7) by $U^{(k)} + U^{(k-1)}$ and integrating on Ω. Using the positivity of the bilinear form B in (5.1.6), we get

(5.1.9) $\qquad \delta_t^{-1} \left[\int_\Omega [U^{(k)}]^2 dx dy \; - \; \int_\Omega [U^{(k-1)}]^2 dx dy \right] \le 0 \qquad k \ge 1,$

which is the discrete counterpart of (5.1.5). A deeper discussion on these issues can be found in DAUTRAY and LIONS (1988-1993), XX, section 2.7.

The scheme (5.1.7) is implicit, which amounts to solving at any step the following boundary-value problem

(5.1.10) $\qquad \begin{cases} L\psi := -\epsilon \Delta \psi + \vec{\beta} \cdot \vec{\nabla} \psi + 2\delta_t^{-1} \psi = f & \text{in } \Omega, \\[2mm] \psi = 0 & \text{on } \partial\Omega. \end{cases}$

The next step is to introduce an approximation with respect to the space variable. Let us assume that $\Omega =]-1, 1[\times]-1, 1[$ and let $n \ge 2$ be an integer. We already studied the discretization of problem (5.1.10) in section 2.3. Setting $f_1 = f_3 = -\epsilon$, $f_2 = 0$, $f_4 = \beta_1$, $f_5 = \beta_2$, $f_6 = 2/\delta_t$, we construct a grid $(\tau_{i,j}^{(n)}, v_{i,j}^{(n)})$, $0 \le i \le n$, $0 \le j \le n$, as explained in section 2.2. Then, at the step $k \ge 0$, we look for a polynomial $q_n^{(k)} \in \mathbf{P}_n^{*,0}$ such that for $k \ge 1$ one has

(5.1.11) $\qquad \delta_t^{-1} \left[q_n^{(k)} - q_n^{(k-1)} \right] + \frac{1}{2} \left[-\epsilon \Delta q_n^{(k)} + \vec{\beta} \cdot \vec{\nabla} q_n^{(k)} \right]$

$\qquad + \frac{1}{2} \left[-\epsilon \Delta q_n^{(k-1)} + \vec{\beta} \cdot \vec{\nabla} q_n^{(k-1)} \right] = 0 \qquad \text{at the nodes } (\tau_{i,j}^{(n)}, v_{i,j}^{(n)}),$

with $1 \le i \le n-1$, $1 \le j \le n-1$. The initial guess $q_n^{(0)}$ interpolates U_0 at the nodes $(\eta_j^{(n)}, \eta_i^{(n)})$, $0 \le i \le n$, $0 \le j \le n$. The main steps to be followed in order to advance the method are:

1) Find the upwind grid $(\tau_{i,j}^{(n)}, v_{i,j}^{(n)})$, $0 \le i \le n$, $0 \le j \le n$, associated with the operator L in (5.1.10);

2) Compute $q_n^{(0)}$ and $Lq_n^{(0)}$ at the nodes $(\eta_j^{(n)}, \eta_i^{(n)})$, $0 \le i \le n$, $0 \le j \le n$, and denote by $r_n^{(0)}$ the polynomial in $\mathbf{P}_n^*$ interpolating $Lq_n^{(0)}$;

3) Compute $r_n^{(k)} := -r_n^{(k-1)} + 4\delta_t^{-1} q_n^{(k-1)}$ for $k \ge 1$ and $k\delta_t \le T$, and find the solution $q_n^{(k)} \in \mathbf{P}_n^{*,0}$ to the problem $Lq_n^{(k)} = r_n^{(k)}$ by collocation at the nodes $(\tau_{i,j}^{(n)}, v_{i,j}^{(n)})$, $1 \le i \le n-1$, $1 \le j \le n-1$;

The resulting polynomial $q_n^{(k)}$, $k \ge 0$, approximates the function U at time $t_k = k\delta_t$. For $k \ge 1$, when finding the solution to the problem $Lq_n^{(k)} = r_n^{(k)}$

at the collocation nodes, computational time can be saved by using $q_n^{(k-1)}$ as initial guess for the Du Fort-Frankel iteration.

FIG. 5.1.1 - *The polynomials* $q_{16}^{(k)}, k = 10m, \ 0 \leq m \leq 8$, *obtained from* (5.1.11) *with* $\delta_t = \frac{1}{80}\pi$, *when* $\epsilon = \frac{1}{100}$ *and* $\vec{\beta} \equiv (y, -x)$.

The algorithm (5.1.11) is also known in the literature as the *Crank-Nicolson method*. We conjecture that (5.1.11) is unconditionally stable (or with very mild limitations on δ_t), second-order accurate in time, and spectrally accurate in space, although a proof of these statements is not yet available.

It is worthwhile remarking that (5.1.11) is appropriate for the treatment of the equation (5.1.1)-(5.1.2) in the finite interval $[0, T]$ with $\delta_t \to 0$. We would not suggest using the same algorithm to find the *steady state* solution $U_\infty(x, y) := \lim_{t \to +\infty} U(x, y, t)$, $(x, y) \in \bar{\Omega}$, by keeping $\delta_t > 0$ fixed and letting $k \to +\infty$. Such an issue has already been investigated in chapter two and more suitable solution techniques have been devised there (i.e. the preconditioned Du Fort-Frankel method with optimal parameters). We also note that the upwind grid associated with the steady state equation (2.1.4) is not the same adopted for the solution of (5.1.10) which depends on δ_t. Therefore, the polynomial $\lim_{k \to +\infty} q_n^{(k)}$, obtained by iterating (5.1.11) for δ_t fixed, is different from that which we would get by collocating the equation (2.1.4).

As far as numerical results are concerned, we find some plots in Fig. 5.1.1. They show the polynomial $q_n^{(k)}$ for different values of k. In the experiment we took $\epsilon = \frac{1}{100}$, $\vec{\beta}(x, y) \equiv (y, -x)$ (which satisfies $\operatorname{div}\vec{\beta} = 0$), $T = \pi$ and

$$(5.1.12) \quad U_0(x, y) := \begin{cases} 16(1 - 4y^2)^2 x^2 (1 - x)^2 & 0 \leq x \leq 1, \ -\frac{1}{2} \leq y \leq \frac{1}{2}, \\ 0 & \text{elsewhere.} \end{cases}$$

Note that U_0 is not a polynomial in $\bar{\Omega}$. The discretization parameters are $n = 16$ and $\delta_t = \frac{1}{80}\pi$. The upwind grid corresponding to L in (5.1.10) turns out to be very similar to the one given in Fig. 2.2.7. The vector field $\vec{\beta}$ recalls the one already considered in (2.4.1). The "bump" individuated by the initial condition rotates around the point $(0, 0)$ in a 180° angle during the time evolution. In fact, the *characteristic lines* relative to the corresponding *hyperbolic* problem ($\epsilon = 0$) are circumferences (see section 6.2). At the same time, some dissipation due to the presence of the Laplace operator in (5.1.1) enlarges the support of the "bump" (note that in this case the steady state U_∞ is zero). The same data of the previous experiment have been used to obtain the plots of Fig. 5.1.2, except that now $\epsilon = \frac{1}{1000}$. The new solution is much less dissipative. The behavior of these approximated solutions does not vary much by increasing n or diminishing δ_t.

Of course, similar results may be reproduced in the case of equations with a *source term* (i.e. when the right-hand side of (5.1.1) is different from zero), or with non-homogeneous boundary conditions. Time-dependent boundary conditions can be enforced by taking into account (1.1.2) or (3.1.1), where the functions on the right-hand sides also depend on $t \in [0, T]$. Dirichlet-type conditions at the time $k\delta_t \leq T$ are directly imposed on the polynomial $q_n^{(k)}$ in (5.1.11).

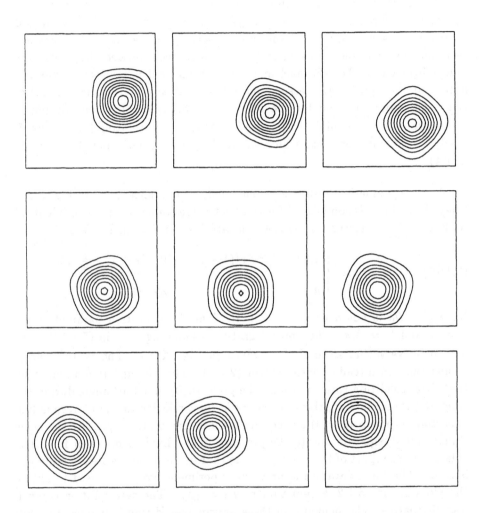

FIG. 5.1.2 - *The polynomials* $q_{16}^{(k)}$, $k = 10m$, $0 \leq m \leq 8$, *obtained from* (5.1.11) *with* $\delta_t = \frac{1}{80}\pi$, *when* $\epsilon = \frac{1}{1000}$ *and* $\vec{\beta} \equiv (y, -x)$.

For Neumann-type conditions in weak form (see section 3.2), there is some more work to be done. For instance, at the nodes on the side $]-1,1[\times\{-1\}$, we generalize (3.2.3) by setting for $k \geq 1$

$$(5.1.13) \qquad \delta_t^{-1}\left[q_n^{(k)} - q_n^{(k-1)}\right](\eta_j^{(n)}, \eta_0^{(n)})$$

$$+ \tfrac{1}{2}\left[(-\epsilon\Delta q_n^{(k)} + \vec{\beta}\cdot\vec{\nabla}q_n^{(k)}) + (-\epsilon\Delta q_n^{(k-1)} + \vec{\beta}\cdot\vec{\nabla}q_n^{(k-1)})\right](\eta_j^{(n)}, \eta_0^{(n)})$$

$$- \frac{\epsilon}{2\tilde{w}_0^{(n)}}\left[\frac{\partial q_n^{(k)}}{\partial y} + \frac{\partial q_n^{(k-1)}}{\partial y}\right](\eta_j^{(n)}\eta_0^{(n)})$$

$$= \frac{\epsilon}{2\tilde{w}_0^{(n)}}\left[h_1(\eta_j^{(n)}, k\delta_t) + h_1(\eta_j^{(n)}, (k-1)\delta_t)\right], \qquad 1 \leq j \leq n-1,$$

$h_1(x,t)$, $x \in]-1,1[$, $t \in [0,T]$, being the Neumann datum. The other three sides and the vertices are similarly handled. Thus, the steady transport-diffusion collocation problem to be solved at any step $k \geq 1$, leads to a linear system including the equations corresponding to the Neumann conditions, which is solved with the usual preconditioned iterative procedure as explained in chapter three.

The adoption of an implicit time-advancing scheme, although it amounts to solving a linear system at each iteration, prevents us from imposing rigid restrictions on the size of the time step. For a full explicit method the maximum δ_t allowed is usually proportional to $\epsilon n^{-4} + |\vec{\beta}|n^{-2}$. Also for ϵ very small, the restriction $\delta_t \approx n^{-2}$ may be very severe, resulting in an expensive implementation. The semi-implicit approach proposed here turns out to be competitive in terms of computational expenses, thanks to the fact that we were able to develop in sections 2.3 and 4.6 a fast solver for transport-dominated equations in the steady case. Higher order multistep finite-differences methods for stiff problems, such as the *BDF methods*, can also be taken into account for the time discretization. These techniques, although more accurate, usually require limitations on the time step.

When Ω has a more complicated geometry, we solve at any iteration the transport-diffusion problem (5.1.10) with the domain decomposition approach described in chapter four. In order to have a second-order accurate scheme in time, the interface conditions have to be carefully imposed. For the sake of simplicity, we only mention the case of the rectangle $\bar{\Omega}$ decomposed into two subdomains, as in Fig. 4.4.1 (see section 4.6), the extension to other geometries being straightforward. We denote by $\hat{\chi}_{n,m}^{(k)}$, $1 \leq m \leq 2$, the two polynomials at the step $k \geq 0$ of the the time discretization process. Inside the square S, $\hat{\chi}_{n,1}^{(k)}$ and $\hat{\chi}_{n,2}^{(k)}$ independently satisfy (5.1.11), where the differential operator

$-\epsilon\Delta + \vec{\beta}\cdot\vec{\nabla}$ is respectively replaced by the operators L_1 and L_2, depending on the geometry of the subdomains Ω_1 and Ω_2 (see (4.6.1)). Recalling (4.6.9)-(4.6.10)-(4.6.11), at the interface we require that for $k \geq 1$

$$
(5.1.14) \qquad \delta_t^{-1}\sigma_1\left[\hat{\chi}_{n,1}^{(k)} - \hat{\chi}_{n,1}^{(k-1)}\right]\tilde{w}_n^{(n)} + \delta_t^{-1}\sigma_2\left[\hat{\chi}_{n,2}^{(k)} - \hat{\chi}_{n,2}^{(k-1)}\right]\tilde{w}_0^{(n)}
$$

$$
+ \tfrac{1}{2}\sigma_1\,(L_1\hat{\chi}_{n,1}^{(k)})(1,\zeta_i^{(n)})\,\tilde{w}_n^{(n)} + \tfrac{1}{2}\sigma_2\,(L_2\hat{\chi}_{n,2}^{(k)})(-1,\zeta_i^{(n)})\,\tilde{w}_0^{(n)}
$$

$$
+ \tfrac{1}{2}\sigma_1\,(L_1\hat{\chi}_{n,1}^{(k-1)})(1,\zeta_i^{(n)})\,\tilde{w}_n^{(n)} + \tfrac{1}{2}\sigma_2\,(L_2\hat{\chi}_{n,2}^{(k-1)})(-1,\zeta_i^{(n)})\,\tilde{w}_0^{(n)}
$$

$$
+ \frac{\epsilon}{8\sigma_1}\frac{\partial\hat{\chi}_{n,1}^{(k)}}{\partial\hat{x}}(1,\zeta_i^{(n)}) - \frac{\epsilon}{8\sigma_2}\frac{\partial\hat{\chi}_{n,2}^{(k)}}{\partial\hat{x}}(-1,\zeta_i^{(n)})
$$

$$
+ \frac{\epsilon}{8\sigma_1}\frac{\partial\hat{\chi}_{n,1}^{(k-1)}}{\partial\hat{x}}(1,\zeta_i^{(n)}) - \frac{\epsilon}{8\sigma_2}\frac{\partial\hat{\chi}_{n,2}^{(k-1)}}{\partial\hat{x}}(-1,\zeta_i^{(n)}) = 0,
$$

with $1 \leq i \leq n - 1$. The collocation nodes are obtained from (4.6.7)-(4.6.8). The equation (5.1.14) is realized, at any step $k \geq 1$, by applying the preconditioned Chebyshev iteration method (4.4.8) up to the number of iterations necessary to reach a prescribed accuracy. At the origin of (5.1.14), there is the imposition in a weak sense of the continuity of the normal derivatives across Γ, with an approach similar to the one followed in (5.1.13) for imposing the Neumann conditions.

We generalize (5.1.1)-(5.1.2)-(5.1.3) by allowing $\vec{\beta}$ to depend on U. The so-obtained nonlinear equation is connected with mathematical models describing various physics phenomena, especially in the case of *conservation laws*, i.e. when $\vec{\beta}(U)\cdot\vec{\nabla}U = \mathrm{div}(\vec{F}(U))$, for a certain vector function $\vec{F}:\mathbf{R}\to\mathbf{R}^2$. We start by observing that the upwind grid determined by the coefficients of the operator in (5.1.10) is now depending on $\vec{\beta}(\psi)$. This means that the collocation nodes need to be updated at any iteration of the time discretization method. The new procedure reads as follows:

1) Define $q_n^{(0)} \in \mathbf{P}_n^{*,0}$ to be the interpolant of U_0 at the nodes $(\eta_j^{(n)},\eta_i^{(n)})$, $0 \leq i \leq n$, $0 \leq j \leq n$, and denote by $r_n^{(0)} \in \mathbf{P}_n^*$ the interpolant of $-\epsilon\Delta q_n^{(0)} + \vec{\beta}(q_n^{(0)})\cdot\vec{\nabla}q_n^{(0)} + 2\delta_t^{-1}q_n^{(0)}$ at the same nodes;

2) For $k \geq 1$ and $k\delta_t \leq T$, construct the upwind grid $(\tau_{i,j}^{(n,k)},v_{i,j}^{(n,k)})$ corresponding to the differential operator $L\psi := -\epsilon\Delta\psi + \vec{\beta}(q_n^{(k-1)})\cdot\vec{\nabla}\psi + 2\delta_t^{-1}\psi$, evaluate $r_n^{(k)} := -r_n^{(k-1)} + 4\delta_t^{-1}q_n^{(k-1)}$ and find the solution $q_n^{(k)} \in \mathbf{P}_n^{*,0}$ to the steady nonlinear equation

$$
-\epsilon\,\Delta q_n^{(k)} + \vec{\beta}(q_n^{(k)})\cdot\vec{\nabla}q_n^{(k)} + 2\delta_t^{-1}q_n^{(k)} = r_n^{(k)},
$$

collocated at the nodes $(\tau_{i,j}^{(n,k)},v_{i,j}^{(n,k)})$, $1 \leq i \leq n - 1$, $1 \leq j \leq n - 1$.

Again, we expect that the polynomial $q_n^{(k)}$, $k \geq 0$, is the approximation of the function U at time $t_k = k\delta_t$. If we intend to choose a small δ_t we may consider, in order to save computational time, to keep the same grid for a certain number of iterations before updating.

To solve the nonlinear equation in step 2 we can adopt a kind of *predictor-corrector* procedure by modifying a little the iterative method introduced in section 2.4. The aim is to construct, for any $k \geq 1$, a sequence of vectors $\{\vec{X}_n^m\}_{m \geq 0}$, representing, at the limit for $m \to +\infty$, the values of the polynomial $q_n^{(k)}$ at the nodes $(\eta_j^{(n)}, \eta_i^{(n)})$, $0 \leq i \leq n$, $0 \leq j \leq n$. We define $\vec{X}_n^0 \equiv \vec{X}_n^1$ to be the vector whose entries are the values of $q_n^{(k-1)}$ at the nodes. At this point, at the iteration $m \geq 1$ of the Du Fort-Frankel method, we update the two coefficients f_4 and f_5 of the discrete operator $\mathcal{L}_n$ using the entries of the vector $\vec{\beta}(\vec{X}_n^m)$. The preconditioning matrix $\mathcal{B}_n$ will not be modified in order to compute its LU factorization only once. If the coefficients of $\mathcal{L}_n$ do not vary too much with m, we expect a convergence behavior similar to that of the linear equation (i.e. when $\vec{\beta}$ is not depending on the solution), even if the preconditioner is supposed to be the optimal one only for the first iteration.

The solutions to nonlinear transport-diffusion equations may develop *internal layers*, generating oscillations in their polynomial approximations. Since the spacing between the grid points is higher at the center of the domain $\bar{\Omega}$ than near the boundary, to get rid of these oscillations one should use a polynomial degree greater than the one used to treat the boundary layers.

We only show a numerical experiment concerning the time evolution in $[0, T]$, $T = 1$, of a kind of two-dimensional *Burgers equation* defined in $\Omega =]-1, 1[\times]-1, 1[$, i.e. when $\vec{\beta}(U) \equiv (U, U)$. We impose homogeneous Dirichlet boundary conditions and the initial guess is $U_0(x, y) = -(1 - y^2)\sin \pi x$, $(x, y) \in \bar{\Omega}$. We have $\epsilon = \frac{1}{25}$, $\delta_t = \frac{1}{40}$ and $n = 16$.
As we can see in Fig. 5.1.3, the two initial "bumps" (the first positive and the other negative) symmetrically move with respect to the center of the square, creating boundary and internal layers. The solution decays, when the time increases, as a result of the high dissipation. A smaller dissipation, realized by diminishing ϵ, requires extra care and a higher degree polynomial resolution.

When we can predict the location of the internal layers, a suitable decomposition of the domain could help improving the numerical results. Since the situations encountered in the approximation of nonlinear problems are so numerous, we are unable to prescribe a general recipe, so that we prefer to avoid comparisons between different strategies. Some tricks were investigated in KOPRIVA (1986), MACARAEG and STREETT (1986), KOPRIVA (1991), and HANLEY (1993). Other comments are provided at the end of section 5.5.

Different kinds of boundary conditions for the spectral approximation of time-dependent advection-diffusion equations are examined in MOFID and PEYRET (1992).

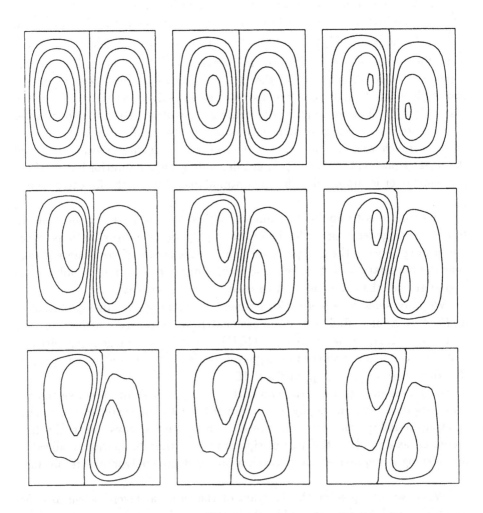

FIG. 5.1.3 - *The polynomials* $q_n^{(k)}$, $k = 5m$, $0 \leq m \leq 8$, *approximating the equation* $\frac{\partial}{\partial t}U - \frac{1}{25}\Delta U + U\frac{\partial}{\partial x}U + U\frac{\partial}{\partial y}U = 0$, *for* $\delta_t = \frac{1}{40}$ *and* $n = 16$.

5.2 The incompressible Navier-Stokes equations

A wide range of applications in *incompressible* fluid dynamics are governed by the *Navier-Stokes* equations (see LANDAU and LIFSCHITZ (1958), BATCHELOR (1967), and MEYER (1971)). In the two-dimensional domain $\bar{\Omega}$, the *velocity field* $\vec{\mathcal{U}} \equiv (U, V) : \bar{\Omega} \times [0, T] \to \mathbf{R}^2$ and the *pressure* $\mathcal{P} : \Omega \times [0, T] \to \mathbf{R}$, have to be determined in order to solve the following time-dependent nonlinear set of equations

(5.2.1)
$$\begin{cases} \frac{\partial}{\partial t}\vec{\mathcal{U}} - \epsilon \, \Delta\vec{\mathcal{U}} + \vec{\mathcal{U}} \cdot \vec{\nabla}\vec{\mathcal{U}} + \vec{\nabla}\mathcal{P} \equiv \vec{F} & \text{in } \Omega \times [0, T], \\[2mm] \vec{\mathcal{U}} \equiv \vec{G} & \text{on } \partial\Omega \times [0, T], \\[2mm] \operatorname{div}\vec{\mathcal{U}} = 0 & \text{in } \Omega \times [0, T], \end{cases}$$

where $\vec{\mathcal{U}} \cdot \vec{\nabla}\vec{\mathcal{U}} := (U\frac{\partial}{\partial x}U + V\frac{\partial}{\partial y}U,\ U\frac{\partial}{\partial x}V + V\frac{\partial}{\partial y}V)$ and $\operatorname{div}\vec{\mathcal{U}} := \frac{\partial}{\partial x}U + \frac{\partial}{\partial y}V$. The *kinematic viscosity* $\epsilon > 0$ is a parameter related to the reciprocal of the so-called *Reynolds number*, and $\vec{F}$ and $\vec{G}$ are given time-dependent vector fields respectively defined on Ω and $\partial\Omega$. Initial conditions are assigned by setting $\vec{\mathcal{U}} \equiv \vec{\mathcal{U}_0}$ in Ω at $t = 0$, with $\operatorname{div}\vec{\mathcal{U}_0} = 0$. The first vector equation in (5.2.1) comes from imposing the conservation of *momentum* for the fluid in motion, and the last scalar equation represents the conservation of *mass*, in the hypothesis of constant *density* (incompressibility). Note that the relation $\operatorname{div}\vec{\mathcal{U}} = 0$ implies $\int_{\partial\Omega}\vec{G}\cdot\vec{\nu} = 0$, $\vec{\nu}$ being a unitary vector field normal to $\partial\Omega$.

A simplified steady linear version of (5.2.1) is the well-known *Stokes problem*, where velocity and pressure (no longer depending on t) are required to satisfy

(5.2.2)
$$\begin{cases} -\epsilon \, \Delta\vec{\mathcal{U}} + \vec{\nabla}\mathcal{P} \equiv \vec{F} & \text{in } \Omega, \\[2mm] \vec{\mathcal{U}} \equiv \vec{G} & \text{on } \partial\Omega, \\[2mm] \operatorname{div}\vec{\mathcal{U}} = 0 & \text{in } \Omega. \end{cases}$$

Some celebrated references concerning the theoretical analysis of Navier-Stokes equations are LADYZHENSKAYA (1969), TEMAM (1985), and KREISS and LORENZ (1989). Here we briefly sketch the proof of existence and uniqueness for problem (5.2.2). Assuming that $\vec{G}$ is zero, we multiply by a vector test function $\vec{\phi}$ vanishing on $\partial\Omega$ the first equation in (5.2.2). The use of the Green formulas yields

(5.2.3) $\qquad \epsilon \displaystyle\int_{\Omega} \vec{\nabla}\vec{\mathcal{U}} \cdot \vec{\nabla}\vec{\phi} \, dx dy - \int_{\Omega} \mathcal{P} \, \operatorname{div}\vec{\phi} \, dx dy = \int_{\Omega} \vec{F} \cdot \vec{\phi} \, dx dy.$

In a suitable *Sobolev* space of functions $\vec{\phi}$ such that $\mathrm{div}\,\vec{\phi} = 0$, the bilinear form on the left-hand side of (5.2.3) is positive definite, providing, with the help of theorem 1.1.2, a unique solution $\vec{\mathcal{U}}$ satisfying the momentum equations in weak form and $\mathrm{div}\,\vec{\mathcal{U}} = 0$. Note that the pressure $\mathcal{P}$, which is not required to satisfy any boundary constraints, is determined up to a constant. From now on, a unique pressure will be determined by setting: $\int_{\Omega}\mathcal{P}\,dxdy = 0$.

For the analysis of the nonlinear case an important role is played by the relation

$$(5.2.4)\qquad \int_{\Omega}(\vec{\mathcal{U}}\cdot\vec{\nabla}\vec{\mathcal{U}})\cdot\vec{\mathcal{U}}\,dxdy \;=\; \tfrac{1}{2}\int_{\Omega}\vec{\mathcal{U}}\cdot\vec{\nabla}|\vec{\mathcal{U}}|^2 dxdy \;=\; -\tfrac{1}{2}\int_{\Omega}|\vec{\mathcal{U}}|^2\mathrm{div}\,\vec{\mathcal{U}}\,dxdy,$$

which brings to the positivity of the operator $\vec{\phi} \to \int_{\Omega}(-\epsilon\Delta\vec{\phi}+\vec{\phi}\cdot\vec{\nabla}\vec{\phi})\cdot\vec{\phi}\,dxdy$ in the space of the functions $\vec{\phi}$ vanishing on $\partial\Omega$ and such that $\mathrm{div}\,\vec{\phi} = 0$.

An idea introduced in BOFFI and FUNARO (1994) provides an alternative formulation of the Navier-Stokes equations. We start by defining a *bubble function* $b : \bar{\Omega} \to \mathbf{R}^{+}$ vanishing on $\partial\Omega$. There is much freedom in choosing the analytic expression of the bubble, provided it preserves a certain qualitative shape. For example, we can require that b is a positive normalized eigenfunction satisfying $-\Delta b = \lambda b$ in Ω, where $\lambda > 0$ is the minimum eigenvalue of the Laplace operator. If $\Omega =]-1,1[\times]-1,1[$, we have $b(x,y) := \cos\tfrac{\pi}{2}x \cos\tfrac{\pi}{2}y$, $(x,y) \in \bar{\Omega}$ (corresponding to $\lambda = \pi^2/2$), but an equivalent choice is also $b(x,y) := (1-x^2)(1-y^2)$, $(x,y) \in \bar{\Omega}$.

Then, for a given positive constant c, we consider the following system in the unknowns $\vec{\mathcal{U}}$ and $\mathcal{P}$:

$$(5.2.5)\qquad \begin{cases} \frac{\partial}{\partial t}\vec{\mathcal{U}} \;-\; \epsilon\,\Delta\vec{\mathcal{U}} \;+\; cb^{-1}\vec{\nabla}b\;\mathrm{div}\vec{\mathcal{U}} \\[2pt] \hspace{3.5cm} +\,\vec{\mathcal{U}}\cdot\vec{\nabla}\vec{\mathcal{U}} \;\equiv\; \vec{F} \;-\; \vec{\nabla}\mathcal{P} \qquad \text{in } \Omega\times[0,T], \\[6pt] \vec{\mathcal{U}} \equiv \vec{G} \hspace{5.5cm} \text{on } \partial\Omega\times[0,T], \\[6pt] -\mathrm{div}(b\vec{\nabla}\mathcal{P}) \;=\; \mathrm{div}\Big(b(\vec{\mathcal{U}}\cdot\vec{\nabla}\vec{\mathcal{U}} \;-\; \vec{F})\Big) \\[4pt] \hspace{2.5cm} +\,\vec{\nabla}b\cdot\Big(\tfrac{\partial}{\partial t}\vec{\mathcal{U}} \;-\; \epsilon\,\Delta\vec{\mathcal{U}}\Big) \qquad \text{in } \Omega\times[0,T]. \end{cases}$$

It is easy to recognize that if $\vec{\mathcal{U}}$ and $\mathcal{P}$ are solutions to (5.2.1), then they also satisfy (5.2.5). In addition, and this is more surprising, the converse is true.

THEOREM 5.2.1 - *There exists a constant $c_0 > 0$ such that, for any $0 < c < c_0$, problem (5.2.5) admits unique solutions $\vec{\mathcal{U}}$ and $\mathcal{P}$ satisfying (5.2.1).*

Proof - We show formally the equivalence of (5.2.5) with (5.2.1) by assuming a sufficient regularity of the solutions. Thus, we set $\mathcal{W} := \mathrm{div}\,\vec{\mathcal{U}}$. Then, multiplying by $\vec{\nabla} b$ the first vector equation in (5.2.5) and substituting in the last equation brings us to

(5.2.6) $$c\,|\vec{\nabla} b|^2 \mathcal{W} \;=\; b^2 \Big[\Delta \mathcal{P} \;+\; \mathrm{div}(\vec{\mathcal{U}} \cdot \vec{\nabla}\vec{\mathcal{U}} \;-\; \vec{F})\Big].$$

Successively, by taking the divergence of the same vector equation with the help of (5.2.6) one gets

(5.2.7)
$$\frac{\partial \mathcal{W}}{\partial t} \;-\; \epsilon \Delta \mathcal{W} \;+\; c\Big[\mathrm{div}(b^{-1}\vec{\nabla} b\;\mathcal{W}) \;+\; b^{-2}|\vec{\nabla} b|^2 \mathcal{W}\Big]$$

$$=\; \frac{\partial \mathcal{W}}{\partial t} \;-\; \epsilon \Delta \mathcal{W} \;+\; cb^{-1}\mathrm{div}(\vec{\nabla} b\;\mathcal{W}) \;=\; 0.$$

If $c > 0$ is sufficiently small (c has to be proportional to ϵ), the equation in (5.2.7) is of parabolic type (see BOFFI and FUNARO (1994) for details). On the other hand, from (5.2.6) we deduce that $\mathcal{W} = 0$ on $\partial\Omega$ since b is also vanishing on the boundary. Finally, having that $\mathcal{W}$ is zero at the time $t = 0$, the unique solution to (5.2.7) in $\bar{\Omega} \times [0, T]$ is also zero. This, together with the condition $\mathrm{div}\,\vec{\mathcal{U}} = 0$, also yields the momentum equation, showing that $\vec{\mathcal{U}}$ and $\mathcal{P}$ satisfy (5.2.1).

In the new formulation proposed, we find in (5.2.5) an elliptic problem of Sturm-Liouville type for the pressure. As we also noticed in section 1.8, this problem does not require boundary conditions for $\mathcal{P}$ due to the positivity of the associated bilinear form (see (1.8.4)). Actually, we have

(5.2.8) $$-\int_\Omega \mathrm{div}(b\,\vec{\nabla}\mathcal{P})\,\mathcal{P}\,dxdy \;=\; \int_\Omega b\,|\vec{\nabla}\mathcal{P}|^2 dxdy \;\geq\; 0,$$

and the equality holds only for $\mathcal{P} = 0$ (recall that $\int_\Omega \mathcal{P}dxdy = 0$). Due to (1.8.3), the integral in Ω of the right-hand side of the pressure equation in (5.2.5) must be zero. This is true, a posteriori, by virtue of the equivalence with (5.2.1).

Reviewing the proof of theorem 5.2.1 in the case of the Stokes problem (5.2.2), we find the system of equations

(5.2.9)
$$\begin{cases} -\,\epsilon\,\Delta\vec{\mathcal{U}} \;+\; cb^{-1}\vec{\nabla} b\,\mathrm{div}\vec{\mathcal{U}} \;\equiv\; \vec{F} \;-\; \vec{\nabla}\mathcal{P} & \text{in } \Omega, \\[2mm] \vec{\mathcal{U}} \;\equiv\; \vec{G} & \text{on } \partial\Omega, \\[2mm] -\mathrm{div}(b\vec{\nabla}\mathcal{P}) \;=\; -\,\epsilon\,\vec{\nabla} b \cdot \Delta\vec{\mathcal{U}} & \text{in } \Omega. \end{cases}$$

5.3 Approximation of the Navier-Stokes equations

Plenty of literature exists regarding the numerical approximation of (5.2.1). Finite-differences or finite elements are common techniques for approaching the problem, but spectral methods, especially in the case of high Reynolds numbers, are preferred by many users. One of the advantages in computing with spectral methods is having a very large spectrum of frequencies to deal with *substructures* at various scales of magnitude generated in the evolution of the fluid. The books of PEYRET and TAYLOR (1982), and CANUTO, HUSSAINI, QUARTERONI and ZANG (1988), provide a general overview and a detailed list of references on the subject.

   We would like to show a possible way to discretize the Navier-Stokes equations. We start by approximating the Stokes problem. The idea of finding the solutions $\vec{\mathcal{U}}$ and $\mathcal{P}$ of (5.2.2) passing through the equivalent formulation (5.2.9) allows the construction of a numerical scheme generating a sequence of elliptic problems where velocity and pressure are decoupled. We briefly describe this *splitting* technique by setting for simplicity $\vec{F} \equiv 0$ in (5.2.9). Although many other efficient methods are available, the one we are going to present gives us the chance to experiment with the solvers for transport-dominated equations developed in the previous chapters.

   We construct suitable sequences $\{\vec{\mathcal{U}}^{(k)}\}_{k \geq 0}$ and $\{\mathcal{P}^{(k)}\}_{k \geq 1}$, respectively converging to $\vec{\mathcal{U}}$ and $\mathcal{P}$. After defining $\vec{\mathcal{U}}^{(0)} \equiv (U^{(0)}, V^{(0)}) := \vec{\mathcal{U}}_0$ with $\operatorname{div} \vec{\mathcal{U}}_0 = 0$, we start with an implicit step to compute the pressure in Ω:

$$(5.3.1) \qquad -\operatorname{div}\left(b\,\vec{\nabla}\mathcal{P}^{(k+1)}\right) \;=\; -\,\epsilon\,\vec{\nabla}b\cdot\Delta\vec{\mathcal{U}}^{(k)} \;+\; \tfrac{1}{4}\epsilon\int_{\Omega}\vec{\nabla}b\cdot\Delta\vec{\mathcal{U}}^{(k)}\,dx\,dy$$

$$\text{with} \qquad \int_{\Omega}\mathcal{P}^{(k+1)}\,dx\,dy \;=\; 0, \qquad k \geq 0.$$

We observe that the right-hand side of (5.3.1) satisfies (1.8.3).
Then, we proceed with another implicit step to compute in $\bar{\Omega}$ the velocity field $\vec{\mathcal{U}}^{(k+1)} \equiv (U^{(k+1)}, V^{(k+1)})$, $k \geq 0$:

$$(5.3.2) \qquad -\,\epsilon\,\Delta U^{(k+1)} \;+\; cb^{-1}\frac{\partial b}{\partial x}\frac{\partial U^{(k+1)}}{\partial x} \;=\; -\frac{\partial \mathcal{P}^{(k+1)}}{\partial x} \;-\; cb^{-1}\frac{\partial b}{\partial x}\frac{\partial V^{(k)}}{\partial y},$$

$$(5.3.3) \qquad -\,\epsilon\,\Delta V^{(k+1)} \;+\; cb^{-1}\frac{\partial b}{\partial y}\frac{\partial V^{(k+1)}}{\partial y} \;=\; -\frac{\partial \mathcal{P}^{(k+1)}}{\partial y} \;-\; cb^{-1}\frac{\partial b}{\partial y}\frac{\partial U^{(k)}}{\partial x}.$$

We set $(U^{(k+1)}, V^{(k+1)}) \equiv \vec{G}$ on $\partial\Omega$. The two boundary-value problems (5.3.2) and (5.3.3) are decoupled and their solution amounts to inverting the two differential operators $L_1 := -\epsilon\Delta + c\left(b^{-1}\frac{\partial}{\partial x}b\right)\frac{\partial}{\partial x}$ and $L_2 := -\epsilon\Delta + c\left(b^{-1}\frac{\partial}{\partial y}b\right)\frac{\partial}{\partial y}$, respectively. Clearly, when the limits $\lim_{k\to+\infty}\mathcal{U}^{(k)}$ and $\lim_{k\to+\infty}\mathcal{P}^{(k)}$ exist, they must satisfy, via (5.2.9), the set of equations in (5.2.2).

Concerning the approximation by spectral methods, we consider the case of the square $\Omega =]-1,1[\times]-1,1[$. The velocity field is approximated at the step $k \geq 0$ by $\vec{q}_n^{(k)} \equiv (q_{1,n}^{(k)}, q_{2,n}^{(k)}) \in \mathbf{P}_n^\star \times \mathbf{P}_n^\star$ and the pressure by a polynomial $r_n^{(k)} \in \mathbf{P}_{n-1}^\star$. The approximation of the pressure, which depends on n^2 degrees of freedom, is obtained by collocation at the Legendre Gauss nodes as follows

$$(5.3.4) \qquad -\left[\mathrm{div}(b\,\vec{\nabla}r_n^{(k+1)})\right](\xi_j^{(n)}, \xi_i^{(n)}) = \Xi^{(k)}(\xi_j^{(n)}, \xi_i^{(n)}) - \Sigma^{(k)},$$

$$\text{for } 1 \leq i \leq n,\ 1 \leq j \leq n, \qquad \text{with} \quad \int_\Omega r_n^{(k+1)}\,dx\,dy = 0,$$

where we defined:

$$\Xi^{(k)} := -\epsilon\left(\frac{\partial b}{\partial x}\Delta q_{1,n}^{(k)} + \frac{\partial b}{\partial y}\Delta q_{2,n}^{(k)}\right) \quad \text{in } \Omega,$$

$$\Sigma^{(k)} := \tfrac{1}{4}\sum_{l,m=1}^{n}\Xi^{(k)}(\xi_m^{(n)}, \xi_l^{(n)})\,w_l^{(n)}w_m^{(n)}.$$

No additional constraints are necessary to solve the set of n^2 equations in (5.3.4). Actually, we recall that the pressure does not need boundary conditions. An algorithm to compute the solution to (5.3.4) has been developed in section 1.8 when $b(x,y) := (1-x^2)(1-y^2)$, $(x,y) \in \bar{\Omega}$. From now on, we shall use this bubble function for our experiments.

In order to determine the right-hand side of (5.3.4), we first compute $\Delta\vec{q}_n^{(k)}$, then, using the two dimensional counterpart of the formula (A.3.3), we recover the values at the set of points $(\xi_j^{(n)}, \xi_i^{(n)})$, $1 \leq i \leq n$, $1 \leq j \leq n$, with a cost proportional to n^3.

Finally, we evaluate the two approximated components of the velocity by collocating the equations (5.3.2) and (5.3.3) at the upwind grid nodes determined by the two transport-diffusion operators L_1 and L_2, respectively (see section 2.2). The two sets of grid points do not depend on k. As usual, for $k \geq 0$, we compute $q_{1,n}^{(k+1)}$ and $q_{2,n}^{(k+1)}$ by employing the preconditioned Du Fort-Frankel method taking as initial guesses $q_{1,n}^{(k)}$ and $q_{2,n}^{(k)}$, respectively. The boundary conditions are imposed at the Legendre nodes on $\partial\Omega$. The gradient of the polynomial

$r_n^{(k+1)}$, obtained by (5.3.4) at the nodes $(\xi_j^{(n)}, \xi_i^{(n)})$, $1 \le i \le n$, $1 \le j \le n$, is extrapolated at the nodes $(\eta_j^{(n)}, \eta_i^{(n)})$, $0 \le i \le n$, $0 \le j \le n$, in order to construct the right-hand sides of the velocity equations.

The algorithm introduced in section 2.4, in particular step 9, requires that the coefficients of the operators L_1 and L_2 are also defined on $\partial\Omega$ (in fact the matrix S_n, used for the interpolation of the residual from the Legendre grid to the upwind grid, also acts on the residual at the boundary). Since $b^{-1}\frac{\partial}{\partial x}b = -2x/(1 - x^2)$ is singular on $\partial\Omega$, before discretizing (5.3.2), we define on the Legendre grid a function B_n such that

$$(5.3.5) \quad B_n(\eta_j^{(n)}, \eta_i^{(n)}) := \begin{cases} \dfrac{-2\eta_j^{(n)}}{1 - [\eta_j^{(n)}]^2} & \text{if } 0 \le i \le n \text{ and } 1 \le j \le n-1, \\[2ex] n^2 & \text{if } 0 \le i \le n \text{ and } j = 0, \\[2ex] -n^2 & \text{if } 0 \le i \le n \text{ and } j = n. \end{cases}$$

Then, as far as the polynomial $q_{1,n}^{(k+1)}$ is concerned, the following equation

$$(5.3.6) \quad -\epsilon\,\Delta q_{1,n}^{(k+1)} + cB_n\frac{\partial q_{1,n}^{(k+1)}}{\partial x} = -\frac{\partial r^{(k+1)}}{\partial x} - cB_n\frac{\partial q_{2,n}^{(k)}}{\partial y},$$

will be satisfied at the upwind grid $(\tau_{i,j}^{(n)}, v_{i,j}^{(n)})$, $1 \le i \le n-1$, $1 \le j \le n-1$, related to the operator L_1. A similar procedure will be applied for determining the polynomial $q_{2,n}^{(k+1)}$.

We remark that, despite theorem 5.2.1, the divergence of $\lim_{k\to+\infty} \vec{q}_n^{(k)}$ is not vanishing. This is partly due to the error committed with the interpolation between the velocity and the pressure grids. The divergence of the approximated velocity field tends however to zero when $n \to +\infty$.

We present some numerical experiments regarding the well-known *square cavity problem*. We set

$$(5.3.7) \quad \vec{G}(x, y) := \begin{cases} (-(1 - x^2)^2, 0) & \text{if } (x, y) \in]-1, 1[\times \{1\}, \\[2ex] (0, 0) & \text{on the other sides of } \partial\Omega. \end{cases}$$

We note that in the Stokes problem, due to the linearity of the equations involved, the velocity field does not depend on ϵ. Therefore, we take $\epsilon = 1$. Finally, we have to decide the constant $c > 0$, and we suggest taking $c = 4\epsilon$. A reduction of c results in a growth of the divergence of $\lim_{k\to+\infty} \vec{q}_n^{(k)}$ and in a less accurate estimation of the pressure. A larger c leads instead to instability, due to a loss of ellipticity of the differential operators L_1 and L_2. With these parameters we compute approximated solutions for different n. We stop the iterative procedure when $k \geq 1$ is such that

$$(5.3.8) \qquad \left(\int_\Omega \left[(q_{1,n}^{(k)} - q_{1,n}^{(k-1)})^2 + (q_{2,n}^{(k)} - q_{2,n}^{(k-1)})^2 \right] dx\,dy \right)^{\frac{1}{2}} < E,$$

where the error E is prescribed. As initial guess we take $\vec{q}_n^{(0)} \equiv \vec{U}_0$, $n \geq 4$, where, for $(x, y) \in \bar{\Omega}$, we have

$$(5.3.9) \quad \vec{U}_0(x, y) := \left(\tfrac{1}{4}(1 - x^2)^2(1 - 2y - 3y^2), x(1 - x^2)(1 + y)(1 - y^2) \right).$$

Note that $\vec{U}_0 \equiv \vec{G}$ on $\partial\Omega$ and $\mathrm{div}\vec{U}_0 = 0$ in $\bar{\Omega}$.

For various n, we give the number of iterations I_n necessary to satisfy (5.3.8) with $E = 10^{-4}$. For $k = I_n$ we also report the norm of the divergence $D_n := \left(\int_\Omega [\mathrm{div}\vec{q}_n^{(k)}]^2 dx\,dy \right)^{1/2}$ and the value $V_n := q_{1,n}^{(k)}(0,0)$ (note that, due to the symmetry of $\vec{G}$, one has $q_{2,n}^{(k)}(0, y) = 0$, $-1 \leq y \leq 1$). These results are provided in Table 5.3.1. A spectral decay of the divergence can be observed. Moreover, the number of iterations I_n does not grow with n, showing that the algorithm is very well-conditioned.

n	I_n	D_n	V_n
6	14	$.135E - 1$	.1646
8	15	$.341E - 2$	.1653
10	16	$.935E - 3$	.1654
12	15	$.434E - 3$	.1654

TABLE 5.3.1 - *Results for the approximation of the Stokes problem.*

In Fig. 5.3.1 one finds the vector field resulting from our computations for $n = 10$ (the intensity of the field has been scaled to allow a readable graphic output, with the symbol ✕— replacing the arrow ⟶).

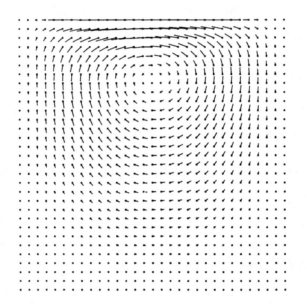

FIG. 5.3.1 - *Velocity vector field for the Stokes problem.*

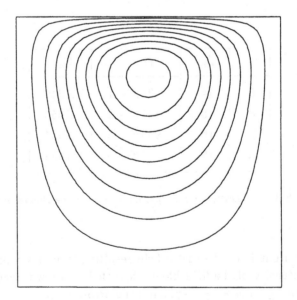

FIG. 5.3.2 - *Stream function for the Stokes problem.*

Because of the condition $\operatorname{div}\vec{\mathcal{U}} = 0$ in Ω, it is easy to recognize that the differential form $d\omega := V\,dx - U\,dy$ is exact. Therefore, there exists a potential function $\Psi : \bar{\Omega} \to \mathbf{R}$, determined up to a constant, such that

$$(5.3.10) \qquad \frac{\partial \Psi}{\partial x} = V \qquad \text{and} \qquad \frac{\partial \Psi}{\partial y} = -U \qquad \text{in } \Omega.$$

Such a function Ψ is called *stream function* (see BATCHELOR (1967), p. 75) and by (5.3.10) its gradient is orthogonal to the vector field $\vec{\mathcal{U}}$. For the square cavity problem, we find that Ψ is constant on $\partial\Omega$. Thus, at the boundary, the following relations are simultaneously satisfied:

$$(5.3.11) \qquad \begin{cases} \Psi = 0 & \text{on } \partial\Omega, \\[2mm] \frac{\partial}{\partial x}\Psi = V & \text{on the sides } \{\pm 1\} \times [-1,1], \\[2mm] \frac{\partial}{\partial y}\Psi = -U & \text{on the sides } [-1,1] \times \{\pm 1\}. \end{cases}$$

Regarding the experiments above, we would also like to provide the plots of the corresponding stream functions (see Fig. 5.3.2 for the case $n = 10$). The numerical determination of Ψ is carried out as follows (see BOFFI and FUNARO (1994)). By differentiating (5.3.10) we get

$$(5.3.12) \qquad -\Delta\Psi = \frac{\partial U}{\partial y} - \frac{\partial V}{\partial x} \qquad \text{in } \Omega,$$

where the right-hand side is known as *vorticity function*.

If we discretize the elliptic problem (5.3.12) in the usual way, we note that there are too many boundary conditions. Thus, we replace (5.2.12) by another equation:

$$(5.3.13) \qquad -b\,\Delta(b^{-1}\Psi) + \left[\frac{2}{b}\left(\frac{\partial b}{\partial x}\right)^2 + \frac{2}{b}\left(\frac{\partial b}{\partial y}\right)^2 - \Delta b \right](b^{-1}\Psi)$$

$$= \frac{\partial U}{\partial y} - \frac{\partial V}{\partial x} + \left(\frac{2}{b}\frac{\partial b}{\partial x}\right)V - \left(\frac{2}{b}\frac{\partial b}{\partial y}\right)U \qquad \text{in } \Omega,$$

where b is the bubble function and the unknown is now $b^{-1}\Psi$.

It is not difficult to check that (5.3.13) is equivalent to (5.3.10) and (5.3.12). Moreover, Ψ is forced by b to be zero on $\partial\Omega$. The other conditions at the boundary deriving from (5.3.11) are imposed by setting

$$
(5.3.14)
\begin{cases}
(b^{-1}\Psi)(x,-1) = \lim_{y\to -1} \dfrac{\Psi(x,y)}{b(x,y)} = -U(x,-1)\left[\tfrac{\partial}{\partial y}b(x,-1)\right]^{-1}, \\[2ex]
(b^{-1}\Psi)(x,1) = \lim_{y\to 1} \dfrac{\Psi(x,y)}{b(x,y)} = -U(x,1)\left[\tfrac{\partial}{\partial y}b(x,1)\right]^{-1}, \\[2ex]
(b^{-1}\Psi)(-1,y) = \lim_{x\to -1} \dfrac{\Psi(x,y)}{b(x,y)} = V(-1,y)\left[\tfrac{\partial}{\partial x}b(-1,y)\right]^{-1}, \\[2ex]
(b^{-1}\Psi)(1,y) = \lim_{x\to 1} \dfrac{\Psi(x,y)}{b(x,y)} = V(1,y)\left[\tfrac{\partial}{\partial x}b(1,y)\right]^{-1}.
\end{cases}
$$

After noting that $\frac{2}{5}\left(\frac{\partial}{\partial x}b\right)^2 + \frac{2}{5}\left(\frac{\partial}{\partial y}b\right)^2 - \Delta b > 0$ in $\bar{\Omega}$, we approximate the solution $b^{-1}\Psi$ to the singular elliptic equation (5.3.13) with the Dirichlet boundary conditions (5.3.14) by the collocation method at the nodes $(\eta_j^{(n)}, \eta_i^{(n)})$, $0 \le i \le n$, $0 \le j \le n$. Note that the right-hand side of (5.3.13) is available at the nodes as a byproduct of the discretization of the Stokes equations. In Fig. 5.3.2 the reader finds the plot corresponding to this computation.

We now discuss the nonlinear case. This time we use (5.2.5). First of all, for $k \ge 0$, we compute the pressure $r_n^{(k+1)} \in \mathbf{P}_{n-1}^*$ with the help of (5.3.4). There are several ways to define the function $\Xi^{(k)}$. We take for example

$$
(5.3.15) \qquad \Xi^{(k)} := \frac{\partial}{\partial x}\left[\tilde{I}_n\left(b\, q_{1,n}^{(k)} \frac{\partial}{\partial x} q_{1,n}^{(k)} + b\, q_{2,n}^{(k)} \frac{\partial}{\partial y} q_{1,n}^{(k)} \right) \right]
$$

$$
+ \frac{\partial}{\partial y}\left[\tilde{I}_n\left(b\, q_{1,n}^{(k)} \frac{\partial}{\partial x} q_{2,n}^{(k)} + b\, q_{2,n}^{(k)} \frac{\partial}{\partial y} q_{2,n}^{(k)} \right) \right] - \epsilon\left(\frac{\partial b}{\partial x}\Delta q_{1,n}^{(k)} + \frac{\partial b}{\partial y}\Delta q_{2,n}^{(k)} \right),
$$

where $\tilde{I}_n : C^0(\bar{\Omega}) \to \mathbf{P}_n^*$ is the interpolation operator at the Legendre grid (see section A.4). To determine the velocity, we first fix the time step $\delta_t > 0$, then modify (5.3.2) and (5.3.3) in the following way

$$
(5.3.16) \qquad \delta_t^{-1}\left[U^{(k+1)} - U^{(k)} \right] - \epsilon\,\Delta U^{(k+1)} + cb^{-1}\frac{\partial b}{\partial x}\frac{\partial U^{(k+1)}}{\partial x}
$$

$$
+ U^{(k)}\frac{\partial U^{(k+1)}}{\partial x} + V^{(k)}\frac{\partial U^{(k+1)}}{\partial y} = -\frac{\partial P^{(k+1)}}{\partial x} - cb^{-1}\frac{\partial b}{\partial x}\frac{\partial V^{(k)}}{\partial y},
$$

(5.3.17) $\qquad \delta_t^{-1}\left[V^{(k+1)} - V^{(k)}\right] - \epsilon \Delta V^{(k+1)} + cb^{-1}\dfrac{\partial b}{\partial y}\dfrac{\partial V^{(k+1)}}{\partial y}$

$$+ U^{(k)}\dfrac{\partial V^{(k+1)}}{\partial x} + V^{(k)}\dfrac{\partial V^{(k+1)}}{\partial y} = -\dfrac{\partial \mathcal{P}^{(k+1)}}{\partial y} - cb^{-1}\dfrac{\partial b}{\partial y}\dfrac{\partial U^{(k)}}{\partial x}.$$

Thus, the two differential operators $L_1 := -\epsilon\Delta + \left(cb^{-1}\frac{\partial}{\partial x}b + U^{(k)}\right)\frac{\partial}{\partial x} + V^{(k)}\frac{\partial}{\partial y} + \delta_t^{-1}$ and $L_2 := -\epsilon\Delta + U^{(k)}\frac{\partial}{\partial x} + \left(cb^{-1}\frac{\partial}{\partial y}b + V^{(k)}\right)\frac{\partial}{\partial y} + \delta_t^{-1}$ are now depending on k. Hence, the upwind grids related to L_1 and L_2 have to be updated at each step. We note that (5.3.16) and (5.3.17) are decoupled as a consequence of the linearization of the advective terms.

The procedure for the computation of the approximated velocity field $\vec{q}_n^{(k+1)}$ follows the standard path already outlined for the linear case. We recall that the numerical implementation requires that L_1 and L_2 are prolongated up to the boundary. In particular, by virtue of (5.3.5), the equation for the first component of the velocity is

(5.3.18) $\qquad \delta_t^{-1}q_{1,n}^{(k+1)} - \epsilon \Delta q_{1,n}^{(k+1)} + cB_n\dfrac{\partial q_{1,n}^{(k+1)}}{\partial x} + q_{1,n}^{(k)}\dfrac{\partial q_{1,n}^{(k+1)}}{\partial x}$

$$+ q_{2,n}^{(k)}\dfrac{\partial q_{1,n}^{(k+1)}}{\partial y} = \delta_t^{-1}q_{1,n}^{(k)} - \dfrac{\partial r^{(k+1)}}{\partial x} - cB_n\dfrac{\partial q_{2,n}^{(k)}}{\partial y}.$$

Similar arguments hold true for the determination of $q_{2,n}^{(k+1)}$.

We computed the solution at the steady state of the cavity problem corresponding to the boundary conditions in (5.3.7). For $\epsilon = \frac{1}{50}$ and $\epsilon = \frac{1}{100}$ we obtained the results of Tables 5.3.2 and 5.3.3, using $c = 4\epsilon$ and $\delta_t = .5$. We took into account the stopping criterion given in (5.3.8) with $E = 10^{-4}$. The quantities I_n, D_n and V_n have the same meaning as those in Table 5.3.1. In Figs. 5.3.3 and 5.3.4 we give the vector fields corresponding to $\epsilon = \frac{1}{50}$ for $n = 12$ and $\epsilon = \frac{1}{100}$ for $n = 16$. The stream functions are given in Figs. 5.3.5 and 5.3.6, respectively. The plots do not substantially change if we increase the degree n. In particular, the location of the *primary vortex* is correct (we remark that, if c is small, the center of the vortex can be misplaced). We found out that the method is unstable when n is too small to resolve the boundary layer that $\vec{\mathcal{U}}$ attains near the side $[-1,1] \times \{1\}$. The scheme needs to be ameliorated since numerical instabilities are also noticed when ϵ is small. On the other hand, the solution of the Navier-Stokes equations for high Reynolds numbers requires special skill, and is beyond the scope of this book in its details.

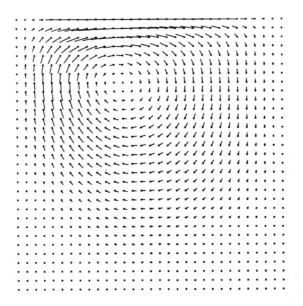

FIG. 5.3.3 - *Velocity field for $\epsilon = \frac{1}{50}$ and $n = 12$.*

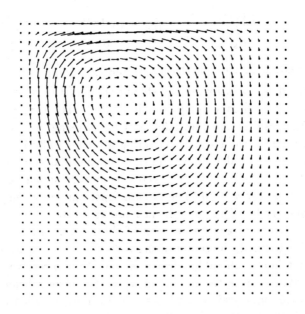

FIG. 5.3.4 - *Velocity field for $\epsilon = \frac{1}{100}$ and $n = 16$.*

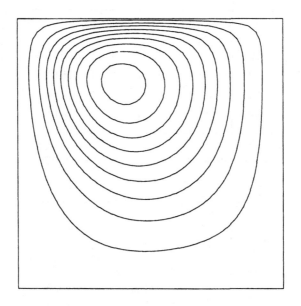

FIG. 5.3.5 - *Stream function for* $\epsilon = \frac{1}{50}$ *and* $n = 12$.

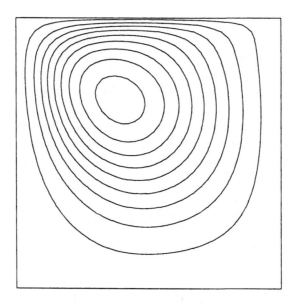

FIG. 5.3.6 - *Stream function for* $\epsilon = \frac{1}{100}$ *and* $n = 16$.

n	I_n	D_n	V_n
10	56	$.100E-1$	.1621
12	56	$.306E-2$	.1612
14	56	$.763E-3$	.1614
16	56	$.391E-3$	.1614

TABLE 5.3.2 - *Results for the nonlinear square cavity problem with* $\epsilon = \frac{1}{50}$.

n	I_n	D_n	V_n
12	91	$.132E-1$	.1576
14	90	$.571E-2$	.1561
16	90	$.152E-2$	.1553
18	90	$.100E-2$	.1554

TABLE 5.3.3 - *Results for the nonlinear square cavity problem with* $\epsilon = \frac{1}{100}$.

Similar results, using the set of Chebyshev collocation nodes and a different numerical approach, were obtained for $\epsilon = \frac{1}{100}$ in MONTIGNY-RANNOU and MORCHOISNE (1987), and for $\epsilon = \frac{1}{50}, \epsilon = \frac{1}{200}$ in EHRENSTEIN and PEYRET (1989). Another discretization scheme, based on formulation (5.2.5) and more suited for the time-dependent case, is experimented in FUNARO, GIANGI and MANSUTTI (1997). Stabilization techniques, via the bubble function approach, for the spectral approximation of the Navier-Stokes equations are examined in CANUTO and VAN KEMENADE (1996).

5.4   The nonlinear Schrödinger equation

We will denote by $i = \sqrt{-1}$ the complex unity. The following partial differential equation

$$(5.4.1) \qquad i\,\frac{\partial U}{\partial t} + \Delta U + |U|^2 U = 0 \quad \text{in } \mathbf{R}^2 \times [0,T], \quad T > 0,$$

arising in quantum mechanics and other physical contexts, is known as the *Schrödinger equation* with cubic nonlinearity (see HASEGAWA (1989)). The unknown $U : \mathbf{R}^2 \times [0,T] \to \mathbf{C}$, attaining complex values, is usually required to decay exponentially at infinity $\forall t \in [0,T]$. An initial condition U_0 at time $t = 0$ is also provided.

By writing $U = V + iW$, where V and W denote the real and imaginary parts of U, respectively, we recover the following system of hyperbolic type

(5.4.2)
$$\begin{cases} \dfrac{\partial V}{\partial t} = -\Delta W - |U|^2 W \\[2mm] \dfrac{\partial W}{\partial t} = \Delta V + |U|^2 V \end{cases} \qquad \text{in } \mathbf{R}^2 \times [0,T].$$

It is also easy to recognize that

(5.4.3)
$$\frac{1}{2}\frac{d}{dt}\int_{\mathbf{R}^2} |U|^2 \, dx\, dy = \int_{\mathbf{R}^2} \left(V \frac{\partial V}{\partial t} + W \frac{\partial W}{\partial t} \right) dx\, dy$$

$$= \int_{\mathbf{R}^2} (-V\,\Delta W + W\,\Delta V)\, dx\, dy$$

$$= \int_{\mathbf{R}^2} \vec{\nabla} V \cdot \vec{\nabla} W \, dx\, dy - \int_{\mathbf{R}^2} \vec{\nabla} W \cdot \vec{\nabla} V \, dx\, dy = 0,$$

which implies that $\int_{\mathbf{R}^2} |U|^2 dx\, dy = \int_{\mathbf{R}^2} |U_0|^2 dx\, dy$ is constant for any $t \in [0,T]$. To obtain (5.4.3) we applied the Green formulas by noticing that V and W are zero at infinity.

By looking for solutions of (5.4.1) of the form

$$U(x,y,t) = \rho(x,y,t) \exp\left[i\left(\tfrac{1}{2}\beta_1 x + \tfrac{1}{2}\beta_2 y + \alpha^2 t - \tfrac{1}{4}(\beta_1^2 + \beta_2^2)t\right)\right],$$

where $\vec{\beta} \equiv (\beta_1, \beta_2) \in \mathbf{R}^2$ and $\alpha \in \mathbf{R}$ are given, we find out that the positive function ρ must satisfy the advection equation:

(5.4.4)
$$\frac{\partial \rho}{\partial t} + \vec{\beta} \cdot \vec{\nabla} \rho = 0 \quad \text{in } \mathbf{R}^2 \times [0,T].$$

From this we deduce that

(5.4.5) $\qquad \rho(x + \beta_1 t, \, y + \beta_2 t, \, t) = \rho_0(x,y), \qquad (x,y) \in \mathbf{R}^2, \quad t \in [0,T],$

i.e. that ρ is constant with respect to t along the straight lines $(x+\beta_1 s, \, y+\beta_2 s)$ with $s \in \mathbf{R}$ (see also section 6.2). The function $\rho_0 : \mathbf{R}^2 \to \mathbf{R}^+$ in (5.4.5) is the so-called *ground state* and turns out to be the solution to the nonlinear elliptic equation

(5.4.6)
$$-\Delta \rho_0 + \alpha^2 \rho_0 - \rho_0^3 = 0 \quad \text{in } \mathbf{R}^2.$$

According to STRAUSS (1978) there exists a radially symmetric positive solution ρ_0 to (5.4.6) decaying to zero at infinity (note that $\rho_0 = 0$ and $\rho_0 = \pm\alpha$ are also solutions).

This means that, if $U_0(x,y) := \rho_0(x,y)\exp[\frac{1}{2}i(\beta_1 x + \beta_2 y)]$, $(x,y) \in \mathbf{R}^2$, the modulus $|U|$ of the solution of the Schrödinger equation is a traveling wave (or *soliton*) shifting in the direction of vector $\vec{\beta}$. The constant α is associated with the amplitude of the soliton.

In order to discretize with respect to variable t, we adopt the implicit scheme

$$
(5.4.7) \quad
\begin{cases}
\delta_t^{-1}[V^{(k)} - V^{(k-1)}] = -\frac{1}{2}[\Delta W^{(k)} + \Delta W^{(k-1)}] \\[2mm]
\qquad\qquad - \frac{1}{4}\Big[|U^{(k)}|^2 + |U^{(k-1)}|^2\Big][W^{(k)} + W^{(k-1)}] \\[2mm]
\delta_t^{-1}[W^{(k)} - W^{(k-1)}] = \frac{1}{2}[\Delta V^{(k)} + \Delta V^{(k-1)}] \\[2mm]
\qquad\qquad + \frac{1}{4}\Big[|U^{(k)}|^2 + |U^{(k-1)}|^2\Big][V^{(k)} + V^{(k-1)}]
\end{cases}
$$

where $\delta_t > 0$ and $k \in \mathbf{N}$ is such that $k\delta_t \leq T$. In (5.4.7), $V^{(k)}$ and $W^{(k)}$ are the real and the imaginary parts of $U^{(k)}$ respectively. We set $U^{(0)} = U_0$. With the same arguments used to obtain (5.4.3), we can prove that the integral $\int_{\mathbf{R}^2} |U^{(k)}|^2 dx dy$ is constant with respect to k.

Most of the results given above are still valid when the equation (5.4.1) is defined on the open subset $\Omega \subset \mathbf{R}^2$ and homogeneous Dirichlet boundary conditions are assumed on $\partial\Omega$. In view of the spectral approximation, we choose the reference square $\Omega =]-1,1[\times]-1,1[$. Successively, for any $k \geq 0$, we approximate $V^{(k)}$ and $W^{(k)}$ in (5.4.7) by the algebraic polynomials $v_n^{(k)} \in \mathbf{P}_n^{\star,0}$ and $w_n^{(k)} \in \mathbf{P}_n^{\star,0}$. We denote by $\vec{X}_n^k$ and $\vec{Y}_n^k$, respectively the vectors of the values attained by $v_n^{(k)}$ and $w_n^{(k)}$ at the Legendre nodes $\Theta_m^{(n)}$, $0 \leq m \leq n_T$, defined in (1.2.3). By collocating the equations in (5.4.7) we obtain in matrix form

$$
(5.4.8) \quad
\begin{pmatrix} 2\delta_t^{-1}\mathcal{I}_n & -\mathcal{A}_n \\[2mm] \mathcal{A}_n & 2\delta_t^{-1}\mathcal{I}_n \end{pmatrix}
\begin{pmatrix} \vec{X}_n^k \\[2mm] \vec{Y}_n^k \end{pmatrix}
$$

$$
= \begin{pmatrix} 2\delta_t^{-1}\mathcal{I}_n & \mathcal{A}_n \\[2mm] -\mathcal{A}_n & 2\delta_t^{-1}\mathcal{I}_n \end{pmatrix}
\begin{pmatrix} \vec{X}_n^{k-1} \\[2mm] \vec{Y}_n^{k-1} \end{pmatrix}
+ \begin{pmatrix} 0 & -\mathcal{T}_n^k \\[2mm] \mathcal{T}_n^k & 0 \end{pmatrix}
\begin{pmatrix} \vec{X}_n^k + \vec{X}_n^{k-1} \\[2mm] \vec{Y}_n^k + \vec{Y}_n^{k-1} \end{pmatrix},
$$

where, as in section 1.2, the matrix $-\mathcal{A}_n$ denotes the Laplace operator in the space $\mathbf{P}_n^{\star,0}$. Moreover, in (5.4.8), $\mathcal{I}_n$ is the $(n+1)^2 \times (n+1)^2$ identity matrix and T_n^k, $k \geq 1$, is the diagonal matrix containing the entries of $\frac{1}{2}([v_n^{(k)}]^2 + [w_n^{(k)}]^2 + [v_n^{(k-1)}]^2 + [w_n^{(k-1)}]^2)$ at the Legendre nodes. Inverting by blocks the matrix on the left-hand side of (5.4.8) (see FUNARO (1992), p. 218), we get

$$
(5.4.9) \quad
\begin{pmatrix} C_n & 0 \\ 0 & C_n \end{pmatrix}
\begin{pmatrix} \vec{X}_n^k \\ \vec{Y}_n^k \end{pmatrix}
=
\begin{pmatrix} 4\delta_t^{-2}I_n - A_n^2 & 4\delta_t^{-1}A_n \\ -4\delta_t^{-1}A_n & 4\delta_t^{-2}I_n - A_n^2 \end{pmatrix}
\begin{pmatrix} \vec{X}_n^{k-1} \\ \vec{Y}_n^{k-1} \end{pmatrix}
$$

$$
+ \begin{pmatrix} 2\delta_t^{-1}I_n & A_n \\ -A_n & 2\delta_t^{-1}I_n \end{pmatrix}
\begin{pmatrix} 0 & -T_n^k \\ T_n^k & 0 \end{pmatrix}
\begin{pmatrix} \vec{X}_n^k + \vec{X}_n^{k-1} \\ \vec{Y}_n^k + \vec{Y}_n^{k-1} \end{pmatrix},
$$

with $C_n := [4\delta_t^{-2}I_n + A_n^2]$.

For any fixed $k \geq 1$, the nonlinear equation (5.4.9) in the unknown $(\vec{X}_n^k, \vec{Y}_n^k)$ can be solved by an iterative procedure by taking as an initial guess the vector $(\vec{X}_n^{k-1}, \vec{Y}_n^{k-1})$ (see DELFOUR, FORTIN and PAYRE (1981)). Actually, each step requires the solution of two linear systems, namely $C_n\vec{X}_n^k = \vec{B}_{1,n}^k$ and $C_n\vec{Y}_n^k = \vec{B}_{2,n}^k$, where the right-hand sides also depend on the unknowns. We use the Du Fort-Frankel method (1.4.6) to carry out both the vectors $\vec{X}_n^k$ and $\vec{Y}_n^k$ at the same time, updating the right-hand sides at each iteration. We also suggest to take as preconditioner the matrix $(2\delta_t^{-1}I_n + B_n)^2$, where $-B_n$ is the finite-differences discretization of the Laplace operator introduced in section 1.4. We observe that we need to factorize the matrix $2\delta_t^{-1}I_n + B_n$ only one time for the entire computation process. A fast convergence is achieved by choosing $\sigma_1 = 1/\sqrt{5}$ and $\sigma_2 = 3/2$ in (1.4.6).

We tried some experiments with the initial Gaussian distribution

$$
U_0(x,y) := \sqrt{\alpha}\, \exp\left[-\alpha((x + \tfrac{1}{2})^2 + y^2) \right]\, \exp\left[\tfrac{1}{2}i(\beta_1(x + \tfrac{1}{2}) + \beta_2 y) \right],
$$

with $(x, y) \in \Omega$, and imposing $U_0 = 0$ on $\partial\Omega$. Due to a result of WEINSTEIN (1983), the inequality

$$
(5.4.10) \quad \int_{\mathbf{R}^2} |U_0|^2\, dx\, dy \; < \; \frac{1}{\alpha} \int_{\mathbf{R}^2} \rho_0^2\, dx\, dy \; \approx \; 1.86,
$$

where ρ_0 is the solution to (5.4.6), guarantees that the solution of (5.4.1) does not explode in a finite time. Our initial condition actually satisfies (5.4.10). By taking $n = 24$, $T = .32$, $\delta_t = T/80$, $\vec{\beta} \equiv (1,2)$ and $\alpha = 15$, we obtain the results of Fig. 5.4.1. The initial configuration, centered at $(-\tfrac{1}{2}, 0)$, is shot in the direction of $\vec{\beta}$ and bounces off the upper side to reach, almost unperturbed, its final position centered at $(\tfrac{1}{2}, 0)$. The plots obtained following the evolution form a sequence of "psychedelic portraits", generated by the disordered recombination of the different *modes* associated with the initial datum, following their own developing rule. These modes magically enter in "resonance" at time T. The phenomenon is discussed with the help of other numerical examples in Yuen and Lake (1978). Further approximations of the Schrödinger equation by spectral methods are presented in Sulem, Sulem and Patera (1984), Bernardi and Pelissier (1994), and De Veronico, Funaro and Reali (1994).

Fig. 5.4.1 - *The approximations* $\sqrt{[v_{24}^{(k)}]^2 + [w_{24}^{(k)}]^2}$, $k = 10m$, $0 \le m \le 8$, *of the modulus* $|U|$ *of the solution to the Schrödinger equation.*

5.5 Semiconductor device equations

We devote this final section to a set of nonlinear equations modelling the movement of *charge carriers* in a semiconductor device, such as a *diode* (see SMITH (1978), SZE (1981), SELBERHERR (1984), and MARKOWICH (1986)). Due to the behavior of the solutions, presenting sharp internal layers, the numerical simulation is a challenge for researchers concerned with transport-dominated equations.

In simplified form, after suitable dimensional scaling, we have to find three functions $\mathcal{P}$, $\mathcal{N}$ and Ψ, defined in $\Omega \subset \mathbf{R}^2$ and depending on time, such that

$$(5.5.1) \qquad -\Delta\Psi = \mathcal{P} - \mathcal{N} + F,$$

$$(5.5.2) \qquad \frac{\partial \mathcal{P}}{\partial t} - \text{div}\left(\epsilon\,\vec{\nabla}\mathcal{P} + \mathcal{P}\,\vec{\nabla}\Psi\right) = 0,$$

$$(5.5.3) \qquad \frac{\partial \mathcal{N}}{\partial t} - \text{div}\left(\epsilon\,\vec{\nabla}\mathcal{N} - \mathcal{N}\,\vec{\nabla}\Psi\right) = 0,$$

where F is a given function and $\epsilon > 0$ is small. The positive unknowns $\mathcal{P}$ and $\mathcal{N}$, respectively denote the concentration of *holes* (of positive charge) and *electrons* (of negative charge). The unknown Ψ is the electric potential, which satisfies the Poisson equation (5.5.1) and is determined up to a constant. Equations (5.5.2) and (5.5.3) are the *continuity equations* for the current densities $\vec{J}_\mathcal{P}$ and $\vec{J}_\mathcal{N}$ given by

$$(5.5.4) \qquad \vec{J}_\mathcal{P} := -\epsilon\,\vec{\nabla}\mathcal{P} - \mathcal{P}\,\vec{\nabla}\Psi, \qquad \vec{J}_\mathcal{N} := \epsilon\,\vec{\nabla}\mathcal{N} - \mathcal{N}\,\vec{\nabla}\Psi.$$

We will be mainly concerned with the determination of the solutions at the steady state. From now on the three unknowns will not depend on time. Before discussing the boundary conditions we show our computational domain (see Fig. 5.5.1).

A *doping* is introduced in the semiconductor crystal represented by the square $\Omega =]-1,1[\times]-1,1[$, so that the borderline γ divides the domain in the positive doped region $\Omega_\mathcal{P}$ and the negative doped region $\Omega_\mathcal{N}$. The function F defined by

$$(5.5.5) \qquad F := \begin{cases} -1 & \text{in } \Omega_\mathcal{P}, \\ 0 & \text{on } \gamma, \\ 1 & \text{in } \Omega_\mathcal{N}, \end{cases}$$

takes into account the doping of the material.

$$\Psi \equiv V \qquad \mathcal{P} \equiv 1 \qquad \mathcal{N} \equiv 0$$

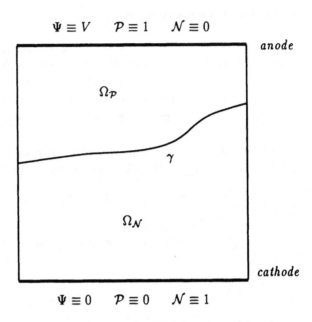

FIG. 5.5.1 - *Design of a* $\mathcal{PN}$*-junction diode.*

The sides $[-1, 1] \times \{\pm 1\}$ (*anode* and *cathode*) are *ohmic contacts* which are forced to assume different potentials. Thus, we require for Ψ the following boundary constraints

(5.5.6)
$$
\begin{cases}
\Psi = 0 & \text{on } [-1, 1] \times \{-1\}, \\[2mm]
\Psi = V & \text{on } [-1, 1] \times \{1\}, \\[2mm]
-\dfrac{\partial \Psi}{\partial x} = 0 & \text{on } \{-1\} \times\,]-1, 1[, \\[2mm]
\dfrac{\partial \Psi}{\partial x} = 0 & \text{on } \{1\} \times\,]-1, 1[,
\end{cases}
$$

where $V \in \mathbf{R}$ is given. The Neumann conditions at the vertical sides say that there the device is electrically insulated (see also section 4.7). Concerning $\mathcal{P}$ and $\mathcal{N}$ we impose

(5.5.7)
$$
\begin{cases}
\mathcal{P} = 0 & \text{on } [-1, 1] \times \{-1\}, \\[2mm]
\mathcal{P} = 1 & \text{on } [-1, 1] \times \{1\}, \\[2mm]
-\epsilon \dfrac{\partial \mathcal{P}}{\partial x} - \mathcal{P} \dfrac{\partial \Psi}{\partial x} = 0 & \text{on } \{-1\} \times\,]-1, 1[, \\[2mm]
\epsilon \dfrac{\partial \mathcal{P}}{\partial x} + \mathcal{P} \dfrac{\partial \Psi}{\partial x} = 0 & \text{on } \{1\} \times\,]-1, 1[,
\end{cases}
$$

$$(5.5.8) \quad \begin{cases} \mathcal{N} = 1 & \text{on } [-1,1] \times \{-1\}, \\[2mm] \mathcal{N} = 0 & \text{on } [-1,1] \times \{1\}, \\[2mm] -\epsilon \dfrac{\partial \mathcal{N}}{\partial x} + \mathcal{N} \dfrac{\partial \Psi}{\partial x} = 0 & \text{on } \{-1\} \times]-1,1[, \\[2mm] \epsilon \dfrac{\partial \mathcal{N}}{\partial x} - \mathcal{N} \dfrac{\partial \Psi}{\partial x} = 0 & \text{on } \{1\} \times]-1,1[. \end{cases}$$

The Dirichlet conditions on the horizontal sides are in agreement with the relation: $\mathcal{P} - \mathcal{N} + F = 0$. The mixed Neumann-Dirichlet conditions at the vertical sides set to zero the normal components of the current densities $\vec{J}_\mathcal{P}$ and $\vec{J}_\mathcal{N}$ introduced in (5.5.4).

In order to decouple the unknowns, we adopt the following iterative algorithm. We start by defining $\Psi^{(0)}(x,y) := \frac{1}{2}V(1+y)$, $(x,y) \in \bar{\Omega}$. Successively, for $k \geq 1$, we advance according to

$$(5.5.9) \qquad -\epsilon \Delta \mathcal{P}^{(k)} - \vec{\nabla}\Psi^{(k-1)} \cdot \vec{\nabla}\mathcal{P}^{(k)} - \Delta \Psi^{(k-1)} \mathcal{P}^{(k)} = 0,$$

$$(5.5.10) \qquad -\epsilon \Delta \mathcal{N}^{(k)} + \vec{\nabla}\Psi^{(k-1)} \cdot \vec{\nabla}\mathcal{N}^{(k)} + \Delta \Psi^{(k-1)} \mathcal{N}^{(k)} = 0,$$

$$(5.5.11) \qquad -\Delta \Psi^{(k)} + \epsilon^{-1}(\mathcal{P}^{(k)} + \mathcal{N}^{(k)})\Psi^{(k)}$$
$$= \mathcal{P}^{(k)} - \mathcal{N}^{(k)} + F + \epsilon^{-1}(\mathcal{P}^{(k)} + \mathcal{N}^{(k)})\Psi^{(k-1)},$$

where for $\Psi^{(k)}$ in (5.5.11) we use the boundary conditions (5.5.6), and for $\mathcal{P}^{(k)}$ and $\mathcal{N}^{(k)}$ we impose (5.5.7) and (5.5.8) where Ψ is replaced by $\Psi^{(k-1)}$. We expect that $\mathcal{P}^{(k)}$, $\mathcal{N}^{(k)}$ and $\Psi^{(k)}$ converge, for $k \to +\infty$, to the steady state solutions of (5.5.1)-(5.5.2)-(5.5.3).

The scheme proposed was first introduced in GUMMEL (1964) (see also MARKOWICH (1986), chapter 3) and is known to perform very well for the kind of differential equations with which we are dealing.

As far as numerical discretization with respect to space variables is concerned, we apply the collocation method at the Legendre nodes for the elliptic equation (5.5.11). The equations (5.5.9) and (5.5.10) are collocated at the upwind grids associated with the operators $-\epsilon\Delta \pm \vec{\nabla}\Psi^{(k-1)} \cdot \vec{\nabla} \pm \Delta \Psi^{(k-1)}$. Details of the implementation are known, so we proceed to the numerical experiments.

We set $\epsilon = \frac{1}{100}$, $n = 28$, and we vary the applied potential V. The $\mathcal{PN}$-junction is described by the curve of equation $\gamma(x) := \frac{3}{16}(1+x)^2$, $x \in [-1,1]$.

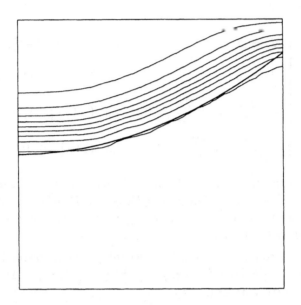

FIG. 5.5.2 - *The function $\mathcal{P}$ for $\epsilon = \frac{1}{100}$ and $V = -.05$.*

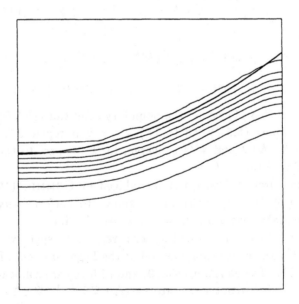

FIG. 5.5.3 - *The function $\mathcal{N}$ for $\epsilon = \frac{1}{100}$ and $V = -.05$.*

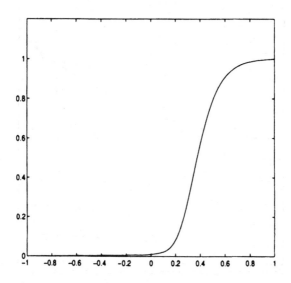

FIG. 5.5.4 - *The function* $\mathcal{P}(0, y)$, $y \in [-1, 1]$, *for* $V = -.05$.

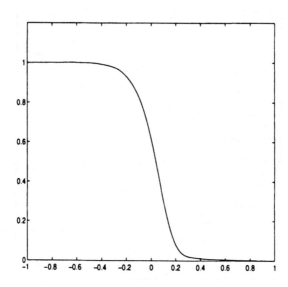

FIG. 5.5.5 - *The function* $\mathcal{N}(0, y)$, $y \in [-1, 1]$, *for* $V = -.05$.

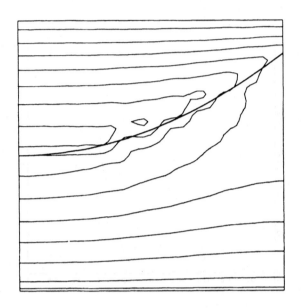

FIG. 5.5.6 - *The function $\mathcal{P}$ for $\epsilon = \frac{1}{100}$ and $V = +.2$.*

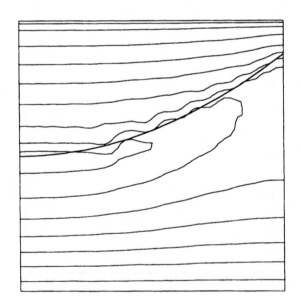

FIG. 5.5.7 - *The function $\mathcal{N}$ for $\epsilon = \frac{1}{100}$ and $V = +.2$.*

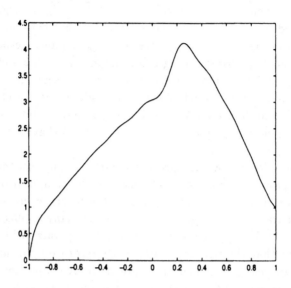

FIG. 5.5.8 - *The function* $\mathcal{P}(0,y)$, $y \in [-1,1]$, *for* $V = +.2$.

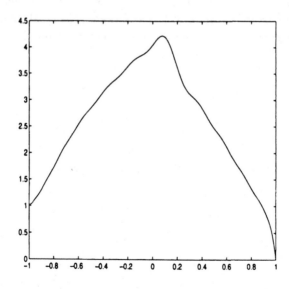

FIG. 5.5.9 - *The function* $\mathcal{N}(0,y)$, $y \in [-1,1]$, *for* $V = +.2$.

If the voltage V is lower than a certain threshold (depending on the doping of the material), then the total current $\vec{J} := \vec{J}_\mathcal{P} + \vec{J}_\mathcal{N}$ is negligible. This situation is called *reverse bias*. Functions $\mathcal{P}$ and $\mathcal{N}$ are between 0 and 1 and present internal layers located in the $\Omega_\mathcal{P}$ and $\Omega_\mathcal{N}$ regions, respectively. For $V = -.05$, we find the corresponding solutions in Figs. 5.5.2 and 5.5.3, while in Figs. 5.5.4 and 5.5.5 we plot the sections along the segment $\{0\} \times [-1, 1]$. If V is larger, the current $\vec{J}$ starts flowing through the device (*forward bias*). The functions $\mathcal{P}$ and $\mathcal{N}$ are allowed to be bigger than 1 and display internal layers along the junction γ. For $V = +.2$ we see the behavior of $\mathcal{P}$ and $\mathcal{N}$ in Figs. 5.5.6, 5.5.7, 5.5.8 and 5.5.9.

Although these results are in agreement with the physics, the approximations look very poor. The degree $n = 28$ used in the computations is still not enough to recover a satisfactory resolution of the layers, even taking into account that the parameter $\epsilon = \frac{1}{100}$ is not as small as many applications require. We can explain this by arguing that the collocation points inside Ω are rare. We are confident that the situation will greatly improve by using the domain decomposition approach described in chapter four. A simple decomposition is $\bar{\Omega} = \bar{\Omega}_\mathcal{P} \cup \bar{\Omega}_\mathcal{N}$ (see section 6.3 for the case of subdomains with bended sides), but the use of more subdomains is strongly recommended since the geometry of a semiconductor device is usually more complicated than the one given in Fig. 5.5.1. Comparisons can be made with the numerous experiments carried out with the help of the finite-differences, finite-boxes or finite element methods. A classical reference is the book of SELBERHERR (1984), although the reader will find more recent results in the numerous papers available.

6
Extensions

We present some other results and open questions related to what has been developed in the previous chapters. The study and improvement of these ideas could be the subject of future investigations.

6.1 A posteriori error estimators

In real life applications there is often the necessity to know *a posteriori* to what extent a certain approximated solution correctly agrees with the exact one. In addition, one is interested to have indications about the discretization parameter in order to calibrate the future computations. This information would be more interesting if one could also localize where the larger discrepancies occurred. Such problems have been already studied for finite element approximations and some references are: BABUSKA and RHEINBOLDT (1978), BANK and WEISER (1985), JOHNSON (1990), AINSWORTH and ODEN (1992), and VERFÜRTH (1994). A posteriori error estimators are also very common in the field of *wavelets* (see MADAY, PERRIER and RAVEL (1991), BERTOLUZZA and NALDI (1994)), allowing a control of the solution before deciding to refine the mesh in some regions of the domain. We are going to see if something similar can be done for approximations using algebraic polynomials in $\Omega =] -1, 1[\times] -1, 1[$.

In chapter two we were able to implement a solver for linear transport-diffusion equations, which guarantees the determination of a global approximating polynomial even if the degree n is not large enough to resolve the boundary layers. The question that arises here concerns the possibility to realize an a posteriori control of the quality of the solution, indicating if and where the choice of n was adequate.

We note that in spectral methods the approximated solution can be evaluated by interpolation and differentiated at any point of the domain $\bar{\Omega}$, so that we can easily obtain the *residual*, which for the approximation of equation (2.1.4) is defined at the Legendre nodes as

$$(6.1.1) \qquad r_n(\eta_j^{(n)}, \eta_i^{(n)}) := \left(-\epsilon \Delta q_n + \vec{\beta} \cdot \vec{\nabla} q_n - f_7 \right)(\eta_j^{(n)}, \eta_i^{(n)}),$$

with $0 \leq i \leq n$, $0 \leq j \leq n$ (see also (1.5.2)). Using the upwind grid approach of section 2.3, we remark that for $\vec{\beta} \not\equiv 0$ the polynomial $r_n \in \mathbf{P}_n^\star$ is required to vanish at a set of nodes not coinciding with the grid $(\eta_j^{(n)}, \eta_i^{(n)})$, $1 \leq i \leq n-1$, $1 \leq j \leq n-1$. We observe that r_n does not vanish at the boundary nodes. For $n \to +\infty$, we find out that r_n converges to zero at the interior of Ω, while, at the boundary $\partial\Omega$, the behavior depends on the source term f_7 which affects the regularity of the exact solution U. For example, in the case of homogeneous Dirichlet boundary conditions and $f_7 = 1$, we obtain $r_n(\pm 1, \pm 1) = -1$, $\forall n \geq 2$.

The analysis of the residual in $\bar{\Omega}$ looks like an efficient way to judge if the polynomial q_n is a correct representation of the exact solution U (see MAVRIPLIS (1989)), especially if the collocation is made at the upwind grid. In fact, it seems that the particular distribution of the upwind nodes amplifies the residual at those regions where the layers are not well resolved. On the other hand, the collocation at the classical Legendre grid leads to oscillating approximated solutions (see FUNARO and RUSSO (1993)) and in contrast the residual is more uniformly distributed. This behavior reminds us of the one shown by the *bubble function* approach of the finite elements, where the discrete solution is *stabilized* by enlarging the approximation space with the help of bubble shaped functions defined in each element. The bubble solution part is the one absorbing the error and indicates the regions where layers not correctly resolved are present. An analysis, specifically devoted to the a posteriori control of the finite element approximations to transport-diffusion equations, is developed in BANK and SMITH (1993), BREZZI and RUSSO (1994), and RUSSO (1996).

Here are a couple of examples. In these experiments we were not interested in knowing about the size of the residual (in fact we did not report any scale in our plots), rather its qualitative distribution in $\bar{\Omega}$. We take $\epsilon = \frac{1}{50}$, $f_7 = 1$ and homogeneous Dirichlet boundary conditions. For $n = 20$ the residuals of two approximated solutions, corresponding to $\vec{\beta} \equiv (1,0)$ and $\vec{\beta} \equiv (1, \frac{1}{2})$, respectively (see Figs. 2.4.3 and 2.4.4), are given in Figs. 6.1.1 and 6.1.2. They are obtained by collocating (2.1.4) at the upwind grid $(\tau_{i,j}^{(n)}, v_{i,j}^{(n)})$, $1 \leq i \leq n-1$, $1 \leq j \leq n-1$, related to the operator $-\epsilon\Delta + \vec{\beta} \cdot \vec{\nabla}$.

FIG. 6.1.1 - *Residuals for* $\epsilon = \frac{1}{50}$, $\vec{\beta} \equiv (1,0)$, $n = 20$,
after collocation at the upwind grid.

FIG. 6.1.2 - *Residuals for* $\epsilon = \frac{1}{50}$, $\vec{\beta} \equiv (1,\frac{1}{2})$, $n = 20$,
after collocation at the upwind grid.

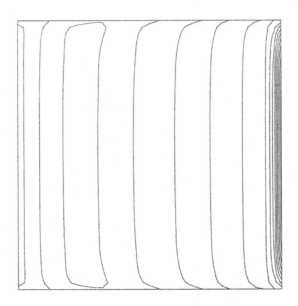

FIG. 6.1.3 - *Residuals for $\epsilon = \frac{1}{50}$, $\vec{\beta} \equiv (1,0)$, $n = 10$,*
after collocation at the upwind grid.

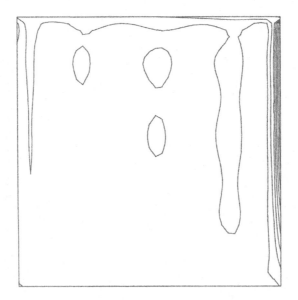

FIG. 6.1.4 - *Residuals for $\epsilon = \frac{1}{50}$, $\vec{\beta} \equiv (1,\frac{1}{2})$, $n = 10$,*
after collocation at the upwind grid.

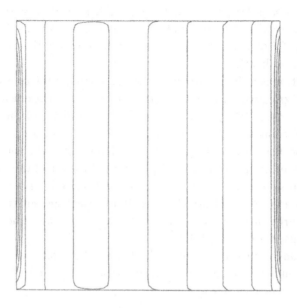

FIG. 6.1.5 - *Residuals for* $\epsilon = \frac{1}{50}$, $\vec{\beta} \equiv (1,0)$, $n = 10$, *after collocation at the Legendre grid.*

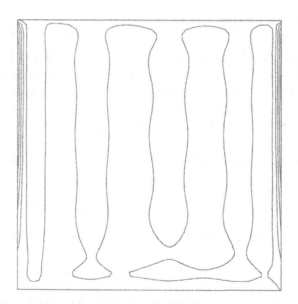

FIG. 6.1.6 - *Residuals for* $\epsilon = \frac{1}{50}$, $\vec{\beta} \equiv (1,\frac{1}{2})$, $n = 10$, *after collocation at the Legendre grid.*

The degree n is large enough to recover a correct representation of the exact solution so that the residuals after collocation at the Legendre nodes $(\eta_j^{(n)}, \eta_i^{(n)})$, $1 \leq i \leq n-1$, $1 \leq j \leq n-1$, are practically the same (hence they are not displayed here).

The situation is different when $n = 10$. In this case the collocation method at the upwind grid gives different results with respect to the one at the Legendre grid (see Figs. 2.4.5, 2.4.6, 2.4.7 and 2.4.8). The residuals of the corresponding approximated solutions are shown in Figs. 6.1.3, 6.1.4, 6.1.5 and 6.1.6. The residual of Fig. 6.1.5 is almost symmetric with respect to the vertical line $x = 0$, therefore, it does not provide indications on the location of the boundary layer. Instead, the residual of Fig. 6.1.3 tells us that something wrong is expected near the side $\{1\} \times]-1, 1[$, and the oscillations at the center of the domain show that a larger n needs to be taken. We arrive at the same conclusions by looking at Figs. 6.1.6 and 6.1.4.

It would be interesting to check whether similar information is available in the case of nonlinear equations, in order to set up a kind of *shock-capturing* technique.

6.2   Pure hyperbolic problems

In this section we examine the discretization of a linear hyperbolic equation. We start by taking $\Omega =]-1, 1[\times]-1, 1[$ and a vector $\vec{\beta} \equiv (\beta_1, \beta_2) : \bar{\Omega} \rightarrow \mathbf{R}^2$. We define the *inflow* boundary $\partial\Omega^- \subset \partial\Omega$ to be the set of boundary points such that $\vec{\beta} \cdot \vec{\nu} < 0$, where $\vec{\nu}$ denotes the outward normal vector to $\partial\Omega$. A corner point V is in $\partial\Omega^-$ if and only if $\vec{\beta} \cdot \vec{\nu} < 0$ for at least one of the two vectors normal to the sides intersecting in V. Similarly, we define the *outflow* boundary as $\partial\Omega^+ = \partial\Omega - \partial\Omega^-$. For example: $V \equiv (1, 1) \in \partial\Omega^+ \Leftrightarrow \beta_1(1, 1) \geq 0$ and $\beta_2(1, 1) \geq 0$.

Then, we are concerned with finding the function $U : \bar{\Omega} \rightarrow \mathbf{R}$ satisfying

$$(6.2.1) \qquad \frac{\partial U}{\partial t} + \vec{\beta} \cdot \vec{\nabla} U = 0 \quad \text{in } \{\Omega \cup \partial\Omega^+\} \times]0, T[,$$

which is obtained from (5.1.1) by setting $\epsilon = 0$. We shall assume that

$$(6.2.2) \qquad U = 0 \quad \text{on } \partial\Omega^- \times [0, T].$$

No boundary conditions are requested on $\partial\Omega^+$. An initial guess $U_0 : \bar{\Omega} \rightarrow \mathbf{R}$, with $U_0 = 0$ on $\partial\Omega^-$, is also assigned at time $t = 0$.

Many results are well-known for the solution U to (6.2.1) (see SMOLLER (1983)). In particular, we define the *characteristic line* $(X(t), Y(t))$, $t \in [0, T]$, passing through $(x, y) \in \Omega$ to be the solution to the system of ordinary differential equations

(6.2.3)
$$
\begin{cases}
\dfrac{d}{dt}X(t) = \beta_1(X(t), Y(t)) \\[2mm]
\dfrac{d}{dt}Y(t) = \beta_2(X(t), Y(t))
\end{cases}
\qquad t \in]0, T],
$$

with $X(0) = x$ and $Y(0) = y$. Then U is constant along the characteristic lines, which means

(6.2.4) $\qquad U(X(t), Y(t), t) = U_0(x, y) \qquad (x, y) \in \Omega, \ t \in [0, T].$

If the flux $\vec{\beta}$ is constant with respect to x and y, the characteristic lines are straight lines and (6.2.4) becomes $U(x + \beta_1 t, y + \beta_2 t, t) = U_0(x, y)$.

Following (5.1.7), an implicit time discretization algorithm is obtained by solving for $k \geq 1$

(6.2.5) $\qquad \delta_t^{-1}\left[U^{(k)} - U^{(k-1)}\right] + \tfrac{1}{2}\vec{\beta} \cdot \vec{\nabla}U^{(k)} + \tfrac{1}{2}\vec{\beta} \cdot \vec{\nabla}U^{(k-1)} = 0,$

with $\delta_t > 0$, $k\delta_t \leq T$. In addition we impose $U^{(0)} = U_0$ in $\bar{\Omega}$ and $U^{(k)} = 0$ on the inflow boundary $\partial\Omega^-$ for $k \geq 0$.

We now turn our attention to space discretization. As in section 2.2, we begin by introducing a suitable set of collocation nodes in $\Omega \cup \partial\Omega^+$. Let $(\eta_j^{(n)}, \eta_i^{(n)})$, $1 \leq i \leq n-1$, $1 \leq j \leq n-1$, be a Legendre node inside Ω, and assume that $|\vec{\beta}|(\eta_j^{(n)}, \eta_i^{(n)}) \neq 0$. Then, we determine the zeroes of the following expressions

(6.2.6) $\qquad 2(1 - [\eta_j^{(n)}]^2)\, P_n'(x^*) = n(n+1)\delta_t\, \beta_1(\eta_j^{(n)}, \eta_i^{(n)})\, P_n(x^*),$

(6.2.7) $\qquad 2(1 - [\eta_i^{(n)}]^2)\, P_n'(y^*) = n(n+1)\delta_t\, \beta_2(\eta_j^{(n)}, \eta_i^{(n)})\, P_n(y^*),$

which are basically (2.2.2) and (2.2.3) when $f_1 = f_2 = f_3 = 0$, $f_4 = \beta_1$, $f_5 = \beta_2$, $f_6 = 2\delta_t^{-1}$. Now, we take ρ as in (2.2.4) to obtain the upwind grid nodes as in (2.2.5), i.e.

(6.2.8) $\qquad \tau_{i,j}^{(n)} := \eta_j^{(n)} - \rho\beta_1(\eta_j^{(n)}, \eta_i^{(n)}), \qquad v_{i,j}^{(n)} := \eta_i^{(n)} - \rho\beta_2(\eta_j^{(n)}, \eta_i^{(n)}).$

Concerning the boundary nodes, we use different arguments depending on whether we are in $\partial\Omega^-$ or $\partial\Omega^+$. First of all, if $(\eta_j^{(n)}, \eta_i^{(n)}) \in \partial\Omega^-$ we use

(6.2.8) with $\rho := 0$. If $(\eta_j^{(n)}, \eta_i^{(n)}) \in \partial\Omega^+$, this point will be shifted in the opposite direction of the flux $\vec{\beta}$, so that there will be no outflow boundary nodes.

For more clarity, we make an example. Let us take the side $\{1\} \times] - 1, 1[$ and assume constant flux $\vec{\beta}$ with $\beta_1 \geq 0$, which yields $(1, \eta_i^{(n)}) \in \partial\Omega^+$ for $1 \leq i \leq n - 1$. Thus, if $\beta_1 > 0$ we look for $\tilde{x}$ such that

$$(6.2.9) \qquad 2\, P'_n(\tilde{x}) \;=\; [n(n+1)]^2 \delta_t \beta_1 \; P_n(\tilde{x}) \qquad \text{with } \tilde{x} > \xi_n^{(n)}.$$

The above equation admits a unique solution. Then, according to FUNARO (1997), one has

$$(6.2.10) \qquad \tilde{x} < 1 \qquad \Leftrightarrow \qquad n(n+1)\delta_t \beta_1 > 1.$$

Hence, we define $x^* := \max\{\tilde{x}, 1\}$ (if $\beta_1 = 0$ we take $x^* := 1$). For $1 \leq i \leq n - 1$, the point $\xi_i^{(n)} < y^* < \xi_{i+1}^{(n)}$ is evaluated with the help of (6.2.7).

Finally, the new nodes $(\tau_{i,n}^{(n)}, v_{i,n}^{(n)})$, $1 \leq i \leq n - 1$, are recovered from (6.2.8) with ρ given in (2.2.4). The treatment of the other sides is similar. One can easily find out how to generalize this procedure in order to determine the upwind nodes corresponding to the corner points in $\partial\Omega^+$, as well as to study the case when $\vec{\beta}$ is not constant on $\bar{\Omega}$. For the sake of simplicity we pass over the details.

Two examples of grids corresponding to the differential operator $\vec{\beta} \cdot \vec{\nabla} + 2\delta_t^{-1}$, are provided in Figs. 6.2.1 and 6.2.2 for $n = 8$ and $\delta_t = .2$. It is not difficult to check that for n fixed and $\delta_t |\vec{\beta}|$ approaching to zero, the upwind grid converges to the usual Legendre grid.

The approximation of $U^{(k)}$, $k \geq 0$, in (6.2.5) will be a polynomial $q_n^{(k)} \in \mathbf{P}_n^*$, $k \geq 0$, satisfying the boundary conditions $q_n^{(k)} = 0$, $k \geq 0$, at the nodes belonging to $\partial\Omega^-$, and such that the differential equation

$$(6.2.11) \qquad 2\delta_t^{-1}\, q_n^{(k)} + \vec{\beta} \cdot \vec{\nabla} q_n^{(k)} = 2\delta_t^{-1}\, q_n^{(k-1)} - \vec{\beta} \cdot \vec{\nabla} q_n^{(k-1)} \qquad k \geq 1,$$

holds true at the remaining upwind nodes. The initial guess $q_n^{(0)}$ is the polynomial interpolating U_0 at $(\eta_j^{(n)}, \eta_i^{(n)})$, $0 \leq i \leq n$, $0 \leq j \leq n$. A slightly different approach, proposed in FUNARO and GOTTLIEB (1988) for the one-dimensional case, consists in collocating the differential equation also at the nodes of $\partial\Omega^-$ and using the boundary conditions as *penalty* terms. Such a technique, which recalls the imposition of the Neumann boundary conditions in weak form for elliptic equations (see section 3.1), will be not considered here.

Any step $k \geq 1$ of (6.2.11) involves the resolution of the linear system $A_n \vec{X}_n = \vec{B}_n$ where the matrix A_n is associated with the spectral discretization of the differential operator $\vec{\beta} \cdot \vec{\nabla} + 2\delta_t^{-1}$ and the unknowns are the values of $q_n^{(k)}$ at the Legendre nodes.

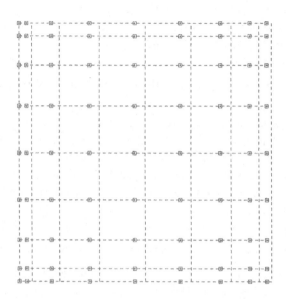

FIG. 6.2.1 - *The upwind grid corresponding to the operator $\frac{\partial}{\partial x} + 10$ for $n = 8$.*

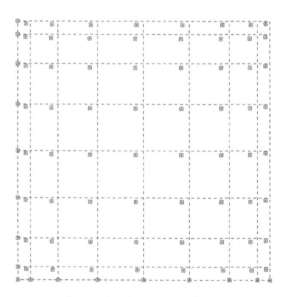

FIG. 6.2.2 - *The upwind grid corresponding to the operator $\frac{\partial}{\partial x} + \frac{1}{2}\frac{\partial}{\partial y} + 10$ for $n = 8$.*

The matrix A_n turns out to be the composition of a spectral differencing operator $\mathcal{L}_n$, obtained by a suitable combination of the entries of the differentiation matrix $\tilde{D}_n$ (see $(A.3.5)$ and Fig. 1.2.2) and the interpolation operator $\mathcal{S}_n$ from the grid $(\eta_j^{(n)} \eta_i^{(n)})$, $0 \le i \le n$, $0 \le j \le n$, to the grid $(\tau_{i,j}^{(n)}, \upsilon_{i,j}^{(n)})$, $0 \le i \le n$, $0 \le j \le n$. Using the notation of section 1.7, we have
$$A_n = \mathcal{J}_n^{\Omega \cup \partial \Omega^+} \mathcal{S}_n \mathcal{L}_n + \mathcal{J}_n^{\partial \Omega^-}.$$

The solution of the linear system can be carried out iteratively using the Du Fort-Frankel method $(1.4.6)$ and the steps mentioned in section 2.4. In order to reduce the costs of implementation we shall perform the interpolation procedure in an approximate way using $(2.2.15)$, although this may affect the accuracy of the method. We finally need to specify the preconditioning matrix B_n. We only propose a preconditioner in the case in which $\vec{\beta}$ is constant on $\bar{\Omega}$ and $\beta_1 > 0$, $\beta_2 > 0$. However, the extension to the general case is not difficult.

For a given vector $\vec{Z}_n$ with entries $\{z_{i,j}\}_{\substack{0 \le i \le n \\ 0 \le j \le n}}$, we specify how to compute $B_n \vec{Z}_n$. For $1 \le i \le n$, $1 \le j \le n$, we consider the first degree polynomial $Q_{i,j} \in \mathbf{P}_1^*$ assuming the values $z_{i,j}$, $z_{i,j+1}$, $z_{i+1,j}$, $z_{i+1,j+1}$ at the nodes $(\eta_j^{(n)}, \eta_i^{(n)})$, $(\eta_{j+1}^{(n)}, \eta_i^{(n)})$, $(\eta_j^{(n)}, \eta_{i+1}^{(n)})$, $(\eta_{j+1}^{(n)}, \eta_{i+1}^{(n)})$, respectively. Applying the differential operator to $Q_{i,j}$ and evaluating at the node $(\tau_{i,j}^{(n)}, \upsilon_{i,j}^{(n)})$ we end up with an upwind finite-difference scheme, i.e.

$$(6.2.12) \qquad \left(\beta_1 \frac{\partial}{\partial x} Q_{i,j} + \beta_2 \frac{\partial}{\partial y} Q_{i,j} + 2\delta_t^{-1} Q_{i,j} \right) (\tau_{i,j}^{(n)}, \upsilon_{i,j}^{(n)})$$

$$= \frac{1}{(\eta_i^{(n)} - \eta_{i-1}^{(n)})(\eta_j^{(n)} - \eta_{j-1}^{(n)})} \Big[\beta_1 \Big((z_{i,j} - z_{i,j-1})(\upsilon_{i,j}^{(n)} - \eta_{i-1}^{(n)})$$

$$+ (z_{i-1,j} - z_{i-1,j-1})(\eta_i^{(n)} - \upsilon_{i,j}^{(n)}) \Big)$$

$$+ \beta_2 \Big((z_{i,j} - z_{i-1,j})(\tau_{i,j}^{(n)} - \eta_{j-1}^{(n)}) + (z_{i,j-1} - z_{i-1,j-1})(\eta_j^{(n)} - \tau_{i,j}^{(n)}) \Big)$$

$$+ 2\delta_t^{-1} \Big(z_{i,j}(\tau_{i,j}^{(n)} - \eta_{j-1}^{(n)})(\upsilon_{i,j}^{(n)} - \eta_{i-1}^{(n)}) - z_{i-1,j}(\tau_{i,j}^{(n)} - \eta_{j-1}^{(n)})(\eta_i^{(n)} - \upsilon_{i,j}^{(n)})$$

$$- z_{i,j-1}(\eta_j^{(n)} - \tau_{i,j}^{(n)})(\upsilon_{i,j}^{(n)} - \eta_{i-1}^{(n)}) + z_{i-1,j-1}(\eta_j^{(n)} - \tau_{i,j}^{(n)})(\eta_i^{(n)} - \upsilon_{i,j}^{(n)}) \Big) \Big].$$

The imposition of the boundary conditions at the nodes on $\partial \Omega^-$ yields $z_{0,j} = 0$, $0 \le j \le n$, and $z_{i,0} = 0$, $1 \le i \le n$. At this point we are able to set up the banded matrix B_n. The structure of B_4 is shown in Fig. 6.2.3.

Similar preconditioners were studied in FUNARO (1987) and FUNARO and ROTHMAN (1989).

```
⎡1 0 0 0 0 0 0 0 0 0 0 0 0 0 0 0 0 0 0 0 0 0 0 0 0⎤
⎢0 1 0 0 0 0 0 0 0 0 0 0 0 0 0 0 0 0 0 0 0 0 0 0 0⎥
⎢0 0 1 0 0 0 0 0 0 0 0 0 0 0 0 0 0 0 0 0 0 0 0 0 0⎥
⎢0 0 0 1 0 0 0 0 0 0 0 0 0 0 0 0 0 0 0 0 0 0 0 0 0⎥
⎢0 0 0 0 1 0 0 0 0 0 0 0 0 0 0 0 0 0 0 0 0 0 0 0 0⎥
⎢0 0 0 0 0 1 0 0 0 0 0 0 0 0 0 0 0 0 0 0 0 0 0 0 0⎥
⎢• • 0 0 0 • • 0 0 0 0 0 0 0 0 0 0 0 0 0 0 0 0 0 0⎥
⎢0 • • 0 0 0 • • 0 0 0 0 0 0 0 0 0 0 0 0 0 0 0 0 0⎥
⎢0 0 • • 0 0 0 • • 0 0 0 0 0 0 0 0 0 0 0 0 0 0 0 0⎥
⎢0 0 0 • • 0 0 0 • • 0 0 0 0 0 0 0 0 0 0 0 0 0 0 0⎥
⎢0 0 0 0 0 0 0 0 0 0 1 0 0 0 0 0 0 0 0 0 0 0 0 0 0⎥
⎢0 0 0 0 0 • • 0 0 0 • • 0 0 0 0 0 0 0 0 0 0 0 0 0⎥
⎢0 0 0 0 0 0 • • 0 0 0 • • 0 0 0 0 0 0 0 0 0 0 0 0⎥
⎢0 0 0 0 0 0 0 • • 0 0 0 • • 0 0 0 0 0 0 0 0 0 0 0⎥
⎢0 0 0 0 0 0 0 0 • • 0 0 0 • • 0 0 0 0 0 0 0 0 0 0⎥
⎢0 0 0 0 0 0 0 0 0 0 0 0 0 0 0 0 1 0 0 0 0 0 0 0 0⎥
⎢0 0 0 0 0 0 0 0 0 0 • • 0 0 0 • • 0 0 0 0 0 0 0 0⎥
⎢0 0 0 0 0 0 0 0 0 0 0 • • 0 0 0 • • 0 0 0 0 0 0 0⎥
⎢0 0 0 0 0 0 0 0 0 0 0 0 • • 0 0 0 • • 0 0 0 0 0 0⎥
⎢0 0 0 0 0 0 0 0 0 0 0 0 0 • • 0 0 0 • • 0 0 0 0 0⎥
⎢0 0 0 0 0 0 0 0 0 0 0 0 0 0 0 0 0 0 0 0 0 1 0 0 0⎥
⎢0 0 0 0 0 0 0 0 0 0 0 0 0 0 0 • • 0 0 0 • • 0 0 0⎥
⎢0 0 0 0 0 0 0 0 0 0 0 0 0 0 0 0 • • 0 0 0 • • 0 0⎥
⎢0 0 0 0 0 0 0 0 0 0 0 0 0 0 0 0 0 • • 0 0 0 • • 0⎥
⎣0 0 0 0 0 0 0 0 0 0 0 0 0 0 0 0 0 0 • • 0 0 0 • •⎦
```

FIG. 6.2.3 - *Structure of the preconditioning matrix $\mathcal{B}_4$*
in the hyperbolic case with $\beta_1 > 0$, $\beta_2 > 0$.

The next step is to check the eigenvalues of the matrix $\mathcal{B}_n^{-1} A_n$. We computed the spectrum of $\mathcal{B}_n^{-1} A_n$ for different choices of n, δ_t and $\vec{\beta}$, obtaining very good results. When the flux $\vec{\beta}$ is parallel to the axes, i.e. $\vec{\beta} \equiv (1,0)$ or $\vec{\beta} \equiv (0,1)$, we are practically reduced to the one-dimensional case, so that we have $n + 1$ groups of equal eigenvalues which are real, positive and between 1 and 1.5 (note that in this case (2.2.15) is equivalent to an exact interpolation in $\mathbf{P}_n^*$). For a general $\vec{\beta}$ the eigenvalues have a positive real part. Most of them are real and the remaining ones have a small imaginary part. They are gathered in a neighborhood around the point 1 of the complex plane, especially when δ_t is small (recall that δ_t is the time step in (6.2.5)).

No substantial growth of the preconditioned condition number is observed by increasing n if δ_t is sufficiently small, say $\delta_t \approx \frac{1}{n}$. The distribution, however, is not too bad even when δ_t is relatively large. We give an example of the distribution of the preconditioned eigenvalues in Fig. 6.2.4, where $n = 8$, $\delta_t = 1$ and $\vec{\beta} \equiv (1, \frac{1}{2})$. It is clear that, with such satisfactory behavior of the spectrum of $B_n^{-1} A_n$, the convergence of the iterative method (1.4.6) is fast, provided a correct choice of the parameters σ_1 and σ_2 is made.

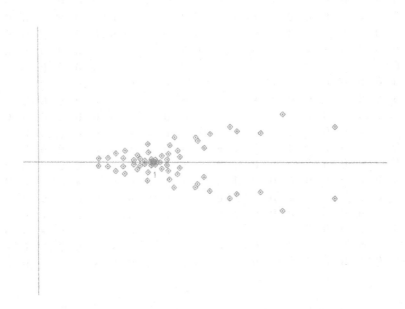

FIG. 6.2.4 - *Eigenvalues of* $B_8^{-1} A_8$ *for* $\delta_t = 1$ *and* $\vec{\beta} \equiv (1, \frac{1}{2})$.

We approximate the problem (6.2.1)-(6.2.2) in the time interval $[0, 2]$ with $U_0(x, y) = [16xy(1 + x)(1 + y)]^2$, $(x, y) \in \bar{\Omega}$, and $\vec{\beta} \equiv (1, \frac{1}{2})$. Due to (6.2.4) the exact solution U shifts in the direction of $\vec{\beta}$. The sequence of polynomials $q_n^{(k)}$, $0 \leq k \leq 80$, obtained for $n = 16$ and $\delta_t = \frac{1}{40}$, evolves in the same manner, as Fig. 6.2.5 shows. The approximated solution shifts at the right velocity and no dissipation is observed. Some small oscillations of size of an order of less that 10% of the size of the moving bump are generated in the interpolation of the initial datum (which is not a polynomial in $\bar{\Omega}$), and can be only detected with a finer graphic resolution.

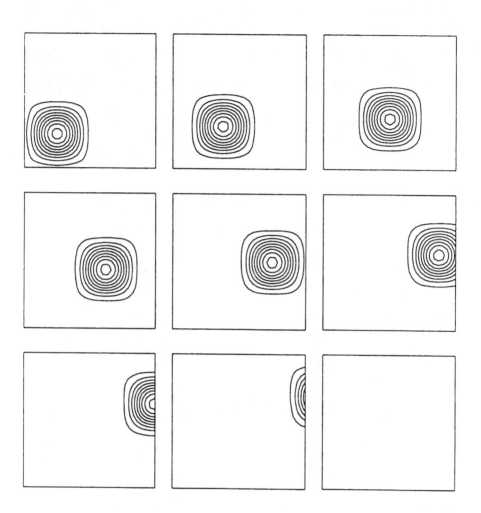

FIG. 6.2.5 - *The polynomials* $q_{16}^{(k)}$, $k = 10m$, $0 \leq m \leq 8$, *obtained from* (6.2.11).

6.3   Elements with bent sides: application to the dam problem

In chapter four we considered the spectral element method in the case of a
conforming decomposition of the domain in a set of quadrilateral subdomains.
In the approximation algorithm discussed, each subdomain was transformed
by isoparametric mapping into the reference square $S =]-1, 1[\times]-1, 1[$, in
order to implement the collocation method. The procedure still works when
the sides of the quadrilaterals, rather than being straight, are slightly bent.

Referring to Fig. 6.3.1, in the subdomain $\bar{\Omega}_m$, the curves related to the sides $\Gamma_k^{(m)}$, $1 \leq k \leq 4$, will be respectively described by the equations

$$
(6.3.1) \quad
\begin{cases}
y = \Upsilon_1(x) & a_1 \leq x \leq a_2 \quad \text{with } b_1 = \Upsilon_1(a_1) \text{ and } b_2 = \Upsilon_1(a_2), \\[2mm]
x = \Upsilon_2(y) & b_2 \leq y \leq b_3 \quad \text{with } a_2 = \Upsilon_2(b_2) \text{ and } a_3 = \Upsilon_2(b_3), \\[2mm]
y = \Upsilon_3(x) & a_4 \leq x \leq a_3 \quad \text{with } b_4 = \Upsilon_3(a_4) \text{ and } b_3 = \Upsilon_3(a_3), \\[2mm]
x = \Upsilon_4(y) & b_1 \leq y \leq b_4 \quad \text{with } a_1 = \Upsilon_4(b_1) \text{ and } a_4 = \Upsilon_4(b_4).
\end{cases}
$$

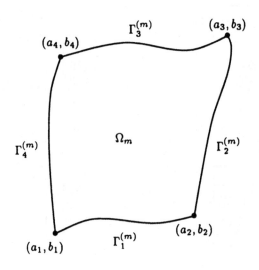

FIG. 6.3.1 - *The domain $\bar{\Omega}_m$ with bent sides.*

Hence, we generalize the mapping $\theta_m \equiv (\theta_{1,m}, \theta_{2,m}) : \bar{S} \to \bar{\Omega}_m$, given in (4.1.1) and (4.1.2), by setting for $-1 \leq \hat{x} \leq 1$, $-1 \leq \hat{y} \leq 1$:

$$
(6.3.2) \quad x = \theta_{1,m}(\hat{x}, \hat{y}) = \tfrac{1}{2}(1 - \hat{x}) \, \Upsilon_4\left(\tfrac{1}{2}(1 - \hat{y})b_1 + \tfrac{1}{2}(1 + \hat{y})b_4\right)
$$

$$
+ \tfrac{1}{2}(1 + \hat{x}) \, \Upsilon_2\left(\tfrac{1}{2}(1 - \hat{y})b_2 + \tfrac{1}{2}(1 + \hat{y})b_3\right),
$$

$$
(6.3.3) \quad y = \theta_{2,m}(\hat{x}, \hat{y}) = \tfrac{1}{2}(1 - \hat{y}) \, \Upsilon_1\left(\tfrac{1}{2}(1 - \hat{x})a_1 + \tfrac{1}{2}(1 + \hat{x})a_2\right)
$$

$$
+ \tfrac{1}{2}(1 + \hat{y}) \, \Upsilon_3\left(\tfrac{1}{2}(1 - \hat{x})a_4 + \tfrac{1}{2}(1 + \hat{x})a_3\right).
$$

Again, a function U defined in Ω_m corresponds to a function $\hat{U} = U(\theta_m)$ defined in S, and we have the relation (4.1.3). Denoting by $\vec{\nu}$ the unitary normal vector to $\Gamma_3^{(m)}$ pointing outside Ω_m, one has

$$(6.3.4) \qquad \frac{\partial U}{\partial \vec{\nu}} = \left(-\Upsilon_3' \frac{\partial U}{\partial x} + \frac{\partial U}{\partial y} \right) \left[1 + (\Upsilon_3')^2 \right]^{-1/2}.$$

Similar relations can be carried out for the other sides.

At this point we can develop, starting from (4.2.13), a spectral element scheme for the solution to the Poisson equation in the set $\bar{\Omega} = \cup_{1 \le m \le M} \bar{\Omega}_m$. The description of the details of such an approximation procedure is quite cumbersome, therefore we proceed no further. For a review of some results on the spectral approximation of equations defined in domains with bent sides, we refer to METIVET (1989), and SCHNEIDESCH and DEVILLE (1994).

We note that in many applications, the analytical expressions of the curves representing the sides are not known, hence, an interpolation at the boundaries is needed. As an example we describe how to handle a *free boundary* problem defined on a single domain, where the upper side (the free boundary) is one of the unknowns. Other *free surface* problems are treated in GARBA, MOFID and PEYRET (1994) and HO and RØNQUIST (1994).

We look for a decreasing function $\Upsilon : [0, 2] \to \mathbf{R}$, with $\Upsilon(0) = 2$ and $\Upsilon(2) > 1$, such that, introducing the domain

$$(6.3.5) \qquad \Omega_\Upsilon := \{(x, y) \mid 0 < x < 2, \ 0 < y < \Upsilon(x)\},$$

the function $U : \bar{\Omega}_\Upsilon \to \mathbf{R}$ satisfies the following boundary-value problem:

$$(6.3.6) \qquad -\Delta U = 0 \qquad \text{in } \Omega_\Upsilon,$$

$$(6.3.7) \qquad \begin{cases} U(0, y) = 2 & 0 \le y \le 2, \\ U(2, y) = 1 & 0 \le y \le 1, \\ U(2, y) = y & 1 < y < \Upsilon(2), \end{cases}$$

$$(6.3.8) \quad -\frac{\partial U}{\partial y}(x, 0) = 0 \quad 0 < x < 2, \quad \text{and} \quad \frac{\partial U}{\partial \vec{\nu}}(x, \Upsilon(x)) = 0 \quad 0 < x \le 2,$$

$$(6.3.9) \qquad U(x, \Upsilon(x)) = \Upsilon(x) \qquad 0 < x \le 2,$$

where $\vec{\nu}$ is the outward normal vector to the upper side. The extra boundary condition (6.3.9) allows the determination of a unique free boundary function Υ.

The above differential problem describes the distribution of water at the steady state filtering across a dam of porous material with a slope from $y = 2$ to $y = 1$ (see Fig. 6.3.2). In particular, the function $U(x, y) - y$, $(x, y) \in \Omega_\Upsilon$, is associated with water pressure due to gravity and Υ is the *line of seepage*. Among the references concerning the physics of the problem we find HARR (1962) and BEAR (1972), while in the book of BAIOCCHI and CAPELO (1984) the reader finds a theoretical analysis of existence and uniqueness. The theory also shows that Υ is concave and that: $\Upsilon'(0) = 0$, $\Upsilon'(2) = -\infty$.

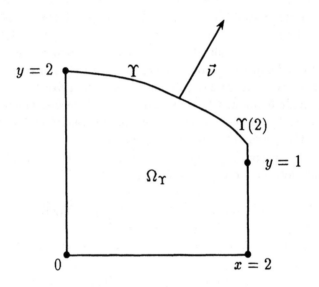

FIG. 6.3.2 - *The domain Ω_Υ in the dam problem.*

The sides of Ω_Υ are determined by the equations

$$(6.3.10) \qquad \Upsilon_1(x) = 0 \quad \text{and} \quad \Upsilon_3(x) = \Upsilon(x) \quad \text{for } 0 \le x \le 2,$$

$$\Upsilon_2(y) = 2, \ 0 \le y \le \Upsilon(2) \quad \text{and} \quad \Upsilon_4(y) = 0, \ 0 \le y \le 2.$$

Then, we use (6.3.2) and (6.3.3) to derive the components of the mapping $\theta \equiv (\theta_1, \theta_2) : \bar{S} \rightarrow \bar{\Omega}_\Upsilon$, i.e.

$$(6.3.11) \ \ \theta_1(\hat{x}, \hat{y}) = 1 + \hat{x}, \qquad \theta_2(\hat{x}, \hat{y}) = \tfrac{1}{2}(1 + \hat{y}) \, \Upsilon(1 + \hat{x}), \qquad (\hat{x}, \hat{y}) \in \bar{S}.$$

Thanks to (4.2.4), the Laplace equation (6.3.6) in the new unknown $\hat{U} = U(\theta)$ becomes

(6.3.12)
$$- \frac{\partial^2 \hat{U}}{\partial \hat{x}^2} + 2(1 + \hat{y}) \frac{\Upsilon'}{\Upsilon} \frac{\partial^2 \hat{U}}{\partial \hat{x} \partial \hat{y}} - \frac{1}{\Upsilon^2} \left(4 + [(1 + \hat{y}) \, \Upsilon']^2 \right) \frac{\partial^2 \hat{U}}{\partial \hat{y}^2}$$

$$+ (1 + \hat{y}) \left(\frac{\Upsilon''}{\Upsilon} - 2 \left(\frac{\Upsilon'}{\Upsilon} \right)^2 \right) \frac{\partial \hat{U}}{\partial \hat{y}} = 0 \quad \text{in } S,$$

Υ, Υ' and Υ'' being evaluated at the point $1 + \hat{x}$.
In a similar way, we transform the boundary conditions:

(6.3.13)
$$\begin{cases} \hat{U}(-1, \hat{y}) = 2 & -1 \leq \hat{y} \leq 1, \\ \hat{U}(1, \hat{y}) = 1 & -1 \leq \hat{y} \leq -1 + 2[\Upsilon(2)]^{-1}, \\ \hat{U}(1, \hat{y}) = \frac{1}{2}(1 + \hat{y})\Upsilon(2) & -1 + 2[\Upsilon(2)]^{-1} < \hat{y} < 1, \end{cases}$$

(6.3.14)
$$- \frac{2}{\Upsilon(1 + \hat{x})} \frac{\partial \hat{U}}{\partial \hat{y}}(\hat{x}, -1) = 0 \qquad -1 < \hat{x} < 1, \quad \text{and}$$

$$- \frac{\Upsilon'(1 + \hat{x})}{\sqrt{1 + [\Upsilon'(1 + \hat{x})]^2}} \frac{\partial \hat{U}}{\partial \hat{x}}(\hat{x}, 1) + \frac{2\sqrt{1 + [\Upsilon'(1 + \hat{x})]^2}}{\Upsilon(1 + \hat{x})} \frac{\partial \hat{U}}{\partial \hat{y}}(\hat{x}, 1) = 0$$

with $-1 < \hat{x} \leq 1$. In the second relation we used (6.3.4). Finally, the counterpart of (6.3.9) is

(6.3.15)
$$\hat{U}(\hat{x}, 1) = \Upsilon(1 + \hat{x}) \qquad -1 < \hat{x} \leq 1.$$

To determine the function Υ we use an iterative approach. We start with a decreasing function $\Upsilon^{(0)}$ such that $\Upsilon^{(0)}(0) = 2$ and $\Upsilon^{(0)}(2) > 1$. Then, we set for $k \geq 1$

(6.3.16)
$$\Upsilon^{(k)}(1 + \hat{x}) = \omega \hat{U}^{(k-1)}(\hat{x}, 1) + (1 - \omega)\Upsilon^{(k-1)}(1 + \hat{x}),$$

where $0 < \omega < 1$ is a relaxation parameter and $\hat{U}^{(k-1)}$ is the solution to (6.3.12) subjected to the boundary conditions given by (6.3.13), (6.3.14), with Υ replaced by $\Upsilon^{(k-1)}$. We expect that $\Upsilon^{(k)}$ converges for $k \to +\infty$ to Υ, since the limit (which does not depend on the initial guess $\Upsilon^{(0)}$) satisfies (6.3.15).

In the discrete case, the function $\hat{U}^{(k)}, k \geq 0$, is approximated by the polynomial $q_n^{(k)} \in \mathbf{P}_n^*$ and the free boundary $\Upsilon^{(k)}, k \geq 0$, by the polynomial $s_n^{(k)} \in \mathbf{P}_n$ in the variable $\hat{x} \in [-1, 1]$. We collocate (6.3.12) as explained in section 2.3, and we impose the boundary conditions as in (6.3.13) and (6.3.14). In particular, for $k \geq 0$, at the corner $(1,1)$ we require that

(6.3.17)
$$-\frac{[s_n^{(k)}]'(1)}{\sqrt{1 + ([s_n^{(k)}]'(1))^2}} \frac{\partial \hat{q}_n^{(k)}}{\partial \hat{x}}(1,1)$$

$$+\frac{2\sqrt{1 + ([s_n^{(k)}]'(1))^2}}{s_n^{(k)}(1)} \frac{\partial \hat{q}_n^{(k)}}{\partial \hat{y}}(1,1) = 0.$$

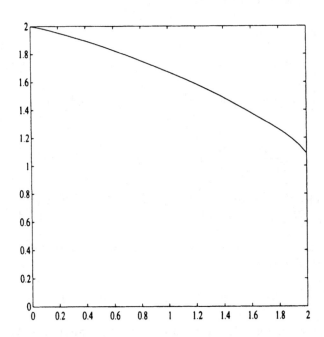

FIG. 6.3.3 - *The free boundary.*

As throughout this book, $q_n^{(k)}$ is determined by a preconditioned iterative solver, this time requiring more iterations than usual due to the condition at the corner $(1,1)$. We observe that the exact solution has a singularity in the derivatives at the point $(2, \Upsilon(2))$, as well as at the point $(2,1)$. The map θ in (6.3.11) is also singular in $(1,1)$. Nevertheless, the numerical results seem to be consistent with those previously obtained by other authors with different techniques (see BAIOCCHI, COMINCIOLI, GUERRI and VOLPI (1973), ALT (1980), BRUCH and SLOSS (1989), and MARINI and PIETRA (1986)). We give in Fig. 6.3.3 the shape of the approximated free boundary obtained for $n = 8$.

6.4   The Poisson equation in 3-D

The approximation techniques developed in chapter one, as well as the extensions provided in successive chapters, can also be applied to problems in three space variables. The generalization is straightforward, however, the cost in memory and computer time drastically grows.

In this section, we adapt the strategy developed for the square in 2-D to approximate the Poisson equation defined in a cube. Therefore, for a given function $f : \Omega \to \mathbf{R}$, we are concerned with finding $U : \bar{\Omega} \to \mathbf{R}$ such that

$$(6.4.1) \quad -\Delta U := -\left(\frac{\partial^2 U}{\partial x^2} + \frac{\partial^2 U}{\partial y^2} + \frac{\partial^2 U}{\partial z^2} \right) = f \quad \text{in } \Omega = (]-1, 1[)^3.$$

Homogeneous Dirichlet boundary conditions are imposed on $\partial\Omega$. For any $n \geq 2$, the solution U to (6.4.1) is approximated by a polynomial $q_n : \bar{\Omega} \to \mathbf{R}$ of degree less or equal to n in each one of the three variables. Such an approximation is made by collocation. Thus, one requires that the equation $-\Delta q_n = f$ is satisfied at the nodes

$$(6.4.2) \qquad \Theta_m^{(n)} \equiv (\eta_i^{(n)}, \eta_j^{(n)}, \eta_k^{(n)}) \qquad m = k(n+1)^2 + j(n+1) + i$$

with $1 \leq i \leq n-1, 1 \leq j \leq n-1, 1 \leq m \leq n-1$. At the remaining Legendre nodes belonging to the boundary $\partial\Omega$, we impose $q_n = 0$.

The collocation problem is equivalent to a linear system $\mathcal{A}_n \vec{X}_n = \vec{B}_n$, where the matrix $-\mathcal{A}_n$ of dimension $(n+1)^3$ is the discretization matrix $\mathcal{L}_n$ of the Laplace operator (the structure of $\mathcal{L}_2$ is provided in Fig. 6.4.1) modified to take into account the boundary conditions (according to the notation of section 1.7 we have $\mathcal{A}_n = \mathcal{J}_n^\Omega \mathcal{L}_n + \mathcal{J}_n^{\partial\Omega}$). The right-hand side $\vec{B}_n$ contains the values of function f at the nodes inside Ω and is set to zero at the entries corresponding to the boundary nodes. The unknown vector $\vec{X}_n$ gives the values of q_n at all points $\Theta_m^{(n)}$, $0 \leq m \leq n_T := (n+1)^3 - 1$, thereby including the boundary points.

For the numerical solution of the linear system, we use the scheme (1.4.6). The preconditioning matrix $-\mathcal{B}_n$ is based on the finite-differences discretization of the Laplace operator, centered at the collocation nodes. Each row of $\mathcal{B}_n$ has seven entries different from zero. The preconditioned eigenvalues are very well distributed providing a fast convergence of the algorithm by correctly choosing the parameters σ_1 and σ_2 in (1.4.7) (the values given in Table 1.4.1

are also effective for the 3-D case). The decay of the residual is the same as in Table 1.5.1. Matrix $\mathcal{B}_n$ is banded with bandwidth $2n^2 + 4n + 3$. Thus, the factorization of $\mathcal{B}_n$ can be very expensive for large n, even when using the fastest solvers for banded matrices available on the market.

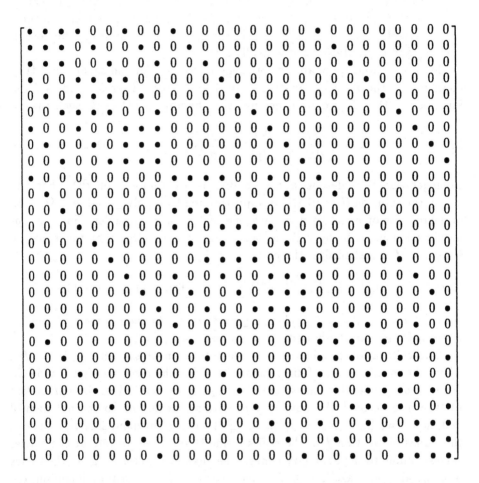

FIG. 6.4.1 - *Structure of the matrix of the Laplace operator in 3-D for $n = 2$.*

This is the bottleneck of our preconditioned iterative procedure. The alternative is to implement the spectral element method by adopting low degree polynomials. We are studying the development of a 3-D code for conforming spectral elements with $n = 5$ (*pentacubes*) based on the upwind grid approach.

The corresponding number of internal unknowns is $4^3 = 64$. A full 64×64 matrix has easy access to each one of the local memories of a parallel computer, where it can be inverted by Gauss elimination, which in this case, is cheaper than using a preconditioned iterative solver. In contrast with the pure spectral approach, a lot of elements are now needed, so that we have to pay more attention to the method applied for determining the solution of the global system associated with the spectral element discretization. The development of a software capable to set up the topological map, taking care of the relations between subdomains (like, for example, determining the common sides, edges and vertices, and controlling the access to the vectors where these interfaces are registered) is quite a complicated task. The software must also organize the outputs, in order to allow a fast use of the graphic environment. The study of the implementation of the pentacubes is in progress and some results will soon be published.

In our opinion, low degree spectral elements will become more and more significant in real life applications, since they can compete well with other classical techniques. Low-order finite elements for example, though very popular, do not provide sufficient accuracy. High-order finite elements (in the h-p version) are difficult to use, especially in 3-D, and, compared to spectral elements, require more degrees of freedom to recover the same accuracy. Finally, pure spectral methods are not so popular and do not perform well in the case of internal layers, as pointed out in section 5.5.

$$\text{———} \quad \diamond \quad \text{———}$$

Our last numerical exercise concerns the discretization of the eikonal equation

$$(6.4.3) \qquad -\epsilon\, \Delta U + |\vec{\nabla} U| = 1 \qquad \text{in } \Omega,$$

already considered in section 1.6 in the 2-D case. We impose homogeneous Dirichlet boundary conditions on $\partial\Omega$. By collocating the equation (6.4.3) at the Legendre grid, we end up with

$$(6.4.4) \qquad \begin{cases} -\epsilon\, \Delta q_n + |\vec{\nabla} q_n| = 1 & \text{at the nodes inside } \Omega, \\[2mm] q_n = 0 & \text{at the boundary nodes,} \end{cases}$$

where q_n is a polynomial of degree n.

For the solution of the nonlinear system (6.4.4), we use a slightly different approach from the one adopted in section 1.6. We compute, via the Du Fort-Frankel scheme, the sequence of vectors

$$(6.4.5) \qquad \vec{X}_n^{k+1} = \frac{2\sigma_1}{1 + 2\sigma_1\sigma_2}\left[(\mathcal{I}_n + \epsilon\delta B_n)^{-1}\left((\mathcal{I}_n + \epsilon\delta A_n)\vec{X}_n^k - \vec{B}_n^k\right)\right.$$

$$\left. + 2\sigma_2\vec{X}_n^k\right] + \frac{1 - 2\sigma_1\sigma_2}{1 + 2\sigma_1\sigma_2}\vec{X}_n^{k-1},$$

representing the values of a sequence of polynomials $\{q_n^{(k)}\}_{k \geq 0}$ at the Legendre nodes. In (6.4.5), $\mathcal{I}_n$ is the identity matrix and $\delta > 0$. The entries of the vector $\vec{B}_n^k$ are now depending on k. They take the values of $q_n^{(k)} + \delta[1 - |\vec{\nabla}q_n^{(k)}|]$ at the nodes inside Ω and are zero in correspondence to the boundary nodes.

We estimate σ_1 and σ_2 using (1.4.9) where $\mu^{(n)}$ is recovered from Table 1.4.1, and we vary δ in order to get the best convergence rate. The iso-level surfaces of Fig. 6.4.1 (compare with Fig. 1.6.2) are obtained for $n = 16$, $\epsilon = .05$ and $\delta = .1$, with about 60 iterations of (6.4.5). The iso-level surfaces of Fig. 6.4.2 (compare with Fig. 1.6.4) are obtained for $n = 20$, $\epsilon = .025$ and $\delta = .06$, in about 110 iterations. In BIANCHI (1995), the reader will find more results, explanations and CPU times.

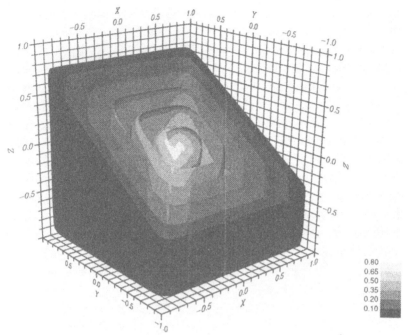

FIG. 6.4.1 - *Approximated solution q_{16} for $\epsilon = \frac{1}{20}$.*

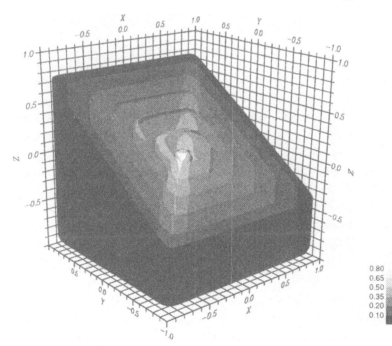

FIG. 6.4.2 - *Approximated solution q_{20} for $\epsilon = \frac{1}{40}$.*

Appendix

We collect here some of the most useful properties of the family of Legendre polynomials P_n, $n \in \mathbf{N}$. More about this topic can be found in the classical texts on orthogonal functions.

A.1 Characterizing properties of Legendre polynomials

The set $\{P_n\}_{n\in\mathbf{N}}$ is a basis in $L^2(-1,1)$ obtained by orthogonalization starting from the basis of monomials $\{x^n\}_{n\in\mathbf{N}}$. Hence, with this definition, one has

$$(A.1.1) \qquad \int_{-1}^{1} P_n(x)P_m(x)\, dx \ = \ 0, \qquad \text{if } n \neq m.$$

We normalize the Legendre polynomials by requiring that $P_n(1) = 1$, $\forall n \in \mathbf{N}$. One easily gets, $\forall x \in \mathbf{R}$:

$$P_0(x) = 1, \quad P_1(x) = x, \quad P_2(x) = \tfrac{3}{2}x^2 - \tfrac{1}{2}, \quad P_3(x) = \tfrac{5}{2}x^3 - \tfrac{3}{2}x,$$

$$P_4(x) = \tfrac{35}{8}x^4 - \tfrac{15}{4}x^2 + \tfrac{3}{8}, \quad \dots\dots$$

and, in general

$$(A.1.2) \qquad P_n(x) \ = \ \frac{1}{2^n\, n!} \sum_{k=0}^{n} (-1)^{n-k} \binom{n}{k} \frac{(2k)!}{(2k-n)!}\, x^{2k-n}.$$

Then, P_n is an odd or even function according to the parity of n. A plot of the Legendre polynomials for $2 \leq n \leq 9$ is given in Fig. A.1.1.

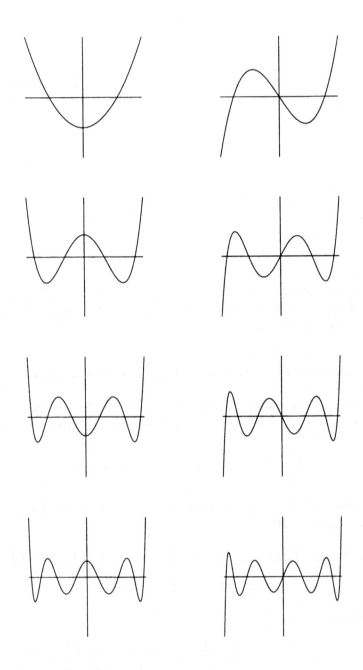

FIG. A.1.1 - *The Legendre polynomials in* $[-1, 1]$ *for* $2 \leq n \leq 9$.

We note that

$$(A.1.3) \qquad \|P_n\|_{L^2(-1,1)}^2 = \int_{-1}^{1} P_n^2(x)\, dx = \frac{2}{2n+1}, \quad \forall n \in \mathbf{N}.$$

Another characterization is given by the formula

$$(A.1.4) \qquad P_n(x) = \frac{1}{2^n\, n!} \frac{d^n}{dx^n}[(x^2 - 1)^n], \quad \forall n \geq 1, \; \forall x \in \mathbf{R}.$$

By $(A.1.4)$ it is straightforward to prove that the Legendre polynomials are solutions of the Sturm-Liouville equation

$$(A.1.5) \quad (1 - x^2)P_n'' - 2x P_n' + n(n+1)P_n = 0, \quad \forall n \in \mathbf{N}, \quad \forall x \in \mathbf{R}.$$

Moreover, $\forall n \geq 2$, $\forall x \in \mathbf{R}$, they satisfy the three-term recursion relation

$$(A.1.6) \qquad P_n(x) = \tfrac{1}{n}[(2n - 1)x P_{n-1}(x) - (n - 1)P_{n-2}(x)].$$

Other useful relations are, $\forall n \in \mathbf{N}$:

$$(A.1.7)\; P_n(\pm 1) = (\pm 1)^n, \quad P_{2n+1}(0) = 0, \quad P_{2n}(0) = (2n)!\, 2^{-2n} \left(\tfrac{1}{n!}\right)^2,$$

$$(A.1.8)\; P_n'(\pm 1) = \pm\tfrac{1}{2}n(n+1)(\pm 1)^n, \quad P_n''(\pm 1) = \tfrac{1}{8}n(n+1)(n^2+n-2)(\pm 1)^n,$$

$$(A.1.9) \qquad |P_n(x)| < 1, \quad |P_n'(x)| < \tfrac{1}{2}n(n + 1), \quad \forall x \in\,] - 1, 1[.$$

A.2 Zeroes and quadrature formulas

It is known that the n-degree Legendre polynomial P_n has n distinct real zeroes in the interval $] - 1, 1[$, which we denote by $\xi_i^{(n)}$, $1 \leq i \leq n$. These zeroes are the nodes of the Gauss integration formula

$$(A.2.1) \qquad \int_{-1}^{1} r\, dx = \sum_{i=1}^{n} r(\xi_i^{(n)})\, w_i^{(n)},$$

which is true for any polynomial r of degree at most $2n - 1$. The weights in $(A.2.1)$ take the following expression:

$$(A.2.2) \qquad w_i^{(n)} = 2 \left[n\, P_n'(\xi_i^{(n)})\, P_{n-1}(\xi_i^{(n)})\right]^{-1}, \qquad 1 \leq i \leq n.$$

FIG. A.2.1 - *Zeroes and extrema of the Legendre polynomials for $2 \leq n \leq 11$.*

The derivative of P_n has $n-1$ distinct real zeroes in $]-1,1[$ which will be denoted by $\eta_i^{(n)}$, $1 \leq i \leq n-1$. After defining $\eta_0^{(n)} = -1$ and $\eta_n^{(n)} = 1$, we obtain another quadrature formula

$$(A.2.3) \qquad \int_{-1}^{1} r \, dx \; = \; \sum_{i=0}^{n} r(\eta_i^{(n)}) \, \tilde{w}_i^{(n)},$$

which is true for any polynomial r of degree at most $2n-1$. The set of the weights in $(A.2.3)$ is given by

$$(A.2.4) \qquad \tilde{w}_i^{(n)} \; = \; \begin{cases} \frac{2}{n(n+1)} & \text{if } i = 0 \text{ or } i = n, \\[2mm] -2[(n+1)P_{n-1}'(\eta_i^{(n)})P_n(\eta_i^{(n)})]^{-1} & \text{if } 1 \leq i \leq n-1. \end{cases}$$

For $2 \leq n \leq 11$, we give in Fig. A.2.1 the distribution of the points $\xi_i^{(n)}$, $1 \leq i \leq n$ (denoted by $\diamond$), and that of the points $\eta_i^{(n)}, 0 \leq i \leq n$ (denoted by $\square$). The distance between two consecutive nodes behaves as n^{-1} at the center of the interval $[-1,1]$, and as n^{-2} near the endpoints.

By virtue of $(A.2.3)$, the Fourier coefficients c_k, $0 \leq k \leq n$, of a polynomial r of degree n, with respect to the basis P_k, $0 \leq k \leq n$, are then obtained by (see FUNARO (1992), p. 67):

$$(A.2.5) \qquad c_k \; = \; \tfrac{1}{2}(2k+1) \int_{-1}^{1} r P_k \, dx$$

$$= \; \tfrac{1}{2}(2k+1) \sum_{i=0}^{n} r(\eta_i^{(n)}) \, P_k(\eta_i^{(n)}) \, \tilde{w}_i^{(n)}, \qquad 0 \leq k \leq n-1,$$

$$(A.2.6) \qquad c_n \; = \; \tfrac{1}{2}(2n+1) \int_{-1}^{1} r P_n \, dx \; = \; \tfrac{1}{2}n \sum_{i=0}^{n} r(\eta_i^{(n)}) \, P_n(\eta_i^{(n)}) \, \tilde{w}_i^{(n)}.$$

Finally, the $L^2(-1,1)$ norm of a polynomial r of degree n can be determined according to (see FUNARO (1992), p. 58):

$$(A.2.7) \qquad \|r\|_{L^2(-1,1)}^2 \; = \; \sum_{k=0}^{n} \frac{2 \, c_k^2}{2k+1}$$

$$= \; \sum_{i=0}^{n} r^2(\eta_i^{(n)}) \, \tilde{w}_i^{(n)} \; - \; \frac{n(n+1)}{2(2n+1)} \sum_{i=0}^{n} r(\eta_i^{(n)}) \, P_n(\eta_i^{(n)}) \, \tilde{w}_i^{(n)}.$$

This also implies

$$(A.2.8) \qquad \int_{-1}^{1} r^2 \, dx \ \leq \ \sum_{i=0}^{n} r^2(\eta_i^{(n)}) \tilde{w}_i^{(n)} \ \leq \ 3 \int_{-1}^{1} r^2 \, dx.$$

A.3   Interpolation and evaluation of derivatives

We can construct the set of n-degree polynomials $\tilde{l}_j^{(n)}$, $0 \leq j \leq n$, with respect to the nodes $\eta_i^{(n)}$, $0 \leq i \leq n$, in such a way that

$$(A.3.1) \qquad \tilde{l}_j^{(n)}(\eta_i^{(n)}) \ = \ \begin{cases} 1 & \text{if } i = j, \\ 0 & \text{if } \neq j. \end{cases}$$

One can show that, for $x \neq \eta_j^{(n)}$, the following expression holds true

$$(A.3.2) \quad \tilde{l}_j^{(n)}(x) = \frac{1}{n(n+1)} \times \begin{cases} -(-1)^n (1-x) P_n'(x) & \text{if } j = 0, \\[2mm] \dfrac{-(1-x^2) P_n'(x)}{P_n(\eta_j^{(n)})(x - \eta_j^{(n)})} & \text{if } 1 \leq j \leq n-1, \\[2mm] (1+x) P_n'(x) & \text{if } j = n. \end{cases}$$

Therefore, for any polynomial r of degree less than or equal to n, we can write

$$(A.3.3) \qquad r(x) \ = \ \sum_{j=0}^{n} r(\eta_j^{(n)}) \, \tilde{l}_j^{(n)}(x), \qquad x \in \mathbf{R}.$$

Deriving the above equation and evaluating at the nodes yields

$$(A.3.4) \qquad r'(\eta_i^{(n)}) \ = \ \sum_{j=0}^{n} \tilde{d}_{ij}^{(1)} \, r(\eta_j^{(n)}), \qquad 0 \leq i \leq n,$$

where $\tilde{d}_{ij}^{(1)} = \left(\frac{d}{dx} \tilde{l}_j^{(n)} \right) (\eta_i^{(n)})$.

The $(n+1) \times (n+1)$ matrix $\tilde{D}_n = \{\tilde{d}_{ij}^{(1)}\}_{\substack{0 \leq i \leq n \\ 0 \leq j \leq n}}$ allows the computation of the derivative of the polynomial r at the nodes, starting from the values attained by the polynomial itself at the nodes. Concerning the entries of $\tilde{D}_n$ we have

$$(A.3.5) \quad \tilde{d}_{ij}^{(1)} = \begin{cases} -\frac{1}{4}n(n+1) & \text{if } i = j = 0, \\[2ex] \dfrac{P_n(\eta_i^{(n)})}{P_n(\eta_j^{(n)})} \dfrac{1}{\eta_i^{(n)} - \eta_j^{(n)}} & \text{if } 0 \le i \le n,\ 0 \le j \le n,\ i \ne j, \\[2ex] 0 & \text{if } 1 \le i = j \le n-1, \\[2ex] \frac{1}{4}n(n+1) & \text{if } i = j = n. \end{cases}$$

The first differentiation matrix $\tilde{D}_n$ is full and non-symmetric. The $(n+1) \times (n+1)$ matrix $\{\tilde{d}_{ij}^{(2)}\}_{\substack{0 \le i \le n \\ 0 \le j \le n}}$, representing the second derivative operator in the space of polynomials of degree less than or equal to n, can be obtained either by squaring $\tilde{D}_n$, or by noting that $\tilde{d}_{ij}^{(2)} = \left(\frac{d^2}{dx^2}\tilde{l}_j^{(n)}\right)(\eta_i^{(n)})$.

Another way to compute the derivative of r is

$$(A.3.6) \qquad r' = \frac{d}{dx}\sum_{k=0}^{n} c_k\, P_k = \sum_{k=0}^{n-1} c_k^{(1)}\, P_k,$$

where the coefficients $c_k^{(1)}$, $0 \le k \le n-1$, can be recovered according to the algorithm

$$(A.3.7) \qquad c_k^{(1)} = (2k+1) \sum_{\substack{i=k+1 \\ k+i\ \text{odd}}}^{n} c_i, \qquad 0 \le k \le n-1.$$

A.4   Approximation results

A function $f \in L^2(-1,1)$ can be approximated by a polynomial expansion as follows:

$$(A.4.1) \qquad f = \sum_{k=0}^{\infty} c_k\, P_k \qquad \text{where}\quad c_k = \frac{2}{2k+1}\int_{-1}^{1} f P_k\, dx, \qquad k \in \mathbf{N}.$$

The projection operator Π_n from $L^2(-1,1)$ to the finite dimensional subspace of polynomials of degree less than or equal to n is defined by

$$(A.4.2) \qquad \Pi_n f = \sum_{k=0}^{n} c_k\, P_k.$$

Then, we have $\lim_{n\to\infty} \|f - \Pi_n f\|_{L^2(-1,1)} = 0$ and the convergence also holds true almost everywhere.

If f is a regular function, we can provide the rate of convergence by the error estimate

$$(A.4.3) \qquad \|f - \Pi_n f\|_{L^2(-1,1)} \leq C\, n^{-\sigma} \left\| (1 - x^2)^{\sigma/2} \frac{d^\sigma f}{dx^\sigma} \right\|_{L^2(-1,1)} \qquad \sigma \geq 0,$$

where the constant $C > 0$ does not depend on n and f.

For a function $f \in C^0([-1,1])$, we can define the interpolation operator $\tilde{I}_n$ in the space of polynomials of degree n, characterized by

$$(A.4.4) \qquad \tilde{I}_n f(\eta_i^{(n)}) = f(\eta_i^{(n)}), \qquad 0 \leq i \leq n.$$

Concerning this operator, BERNARDI and MADAY (1992), p. 77, give the following error estimates

$$(A.4.5) \qquad \|f - \tilde{I}_n f\|_{L^2(-1,1)} \leq C\, n^{-\sigma} \left\| \frac{d^\sigma f}{dx^\sigma} \right\|_{L^2(-1,1)} \qquad \sigma \geq 1,$$

$$(A.4.5) \qquad \|(f - \tilde{I}_n f)'\|_{L^2(-1,1)} \leq C\, n^{1-\sigma} \left\| \frac{d^\sigma f}{dx^\sigma} \right\|_{L^2(-1,1)} \qquad \sigma \geq 1,$$

where $C > 0$ does not depend on n and f.

Finally, we recall the following *inverse inequalities* which hold true for any polynomial r of degree less than or equal to n, with $n \in \mathbf{N}$:

$$(A.4.7) \qquad \|\sqrt{(1 - x^2)}\, r'\|_{L^2(-1,1)} \leq \sqrt{n(n+1)}\, \|r\|_{L^2(-1,1)},$$

$$(A.4.8) \qquad \|r'\|_{L^2(-1,1)} \leq n(n+1)\, \|r\|_{L^2(-1,1)},$$

$$(A.4.9) \qquad \|r'\|_{L^\infty(-1,1)} \leq n^2\, \|r\|_{L^\infty(-1,1)}.$$

A.5   List of symbols

P_n : Legendre polynomial of degree $n \in \mathbf{N}$;

$\xi_i^{(n)}$, $1 \leq i \leq n$: zeroes of P_n, Gauss nodes;

$\eta_i^{(n)}$, $0 \leq i \leq n$: zeroes of $(1 - x^2)P_n'$, Gauss-Lobatto nodes;

$w_i^{(n)}, \quad 1 \le i \le n$ weights of the Gauss quadrature formula;

$\tilde{w}_i^{(n)}, \quad 0 \le i \le n$ weights of the Gauss-Lobatto quadrature formula;

$l_j^{(n)}, \quad 1 \le j \le n$ Lagrange polynomials with respect to the Gauss nodes $\xi_i^{(n)}, \ 1 \le i \le n$;

$\tilde{l}_j^{(n)}, \quad 0 \le j \le n$ Lagrange polynomials with respect to the Gauss-Lobatto nodes $\eta_i^{(n)}, \ 0 \le i \le n$;

$\mathbf{P}_n$ space of polynomials of degree less or equal to n;

$\tilde{D}_n = \{\tilde{d}_{ij}^{(1)}\}_{\substack{0 \le i \le n \\ 0 \le j \le n}}$ derivative matrix in $\mathbf{P}_n$ with respect to the Gauss-Lobatto collocation nodes $\eta_i^{(n)}, \ 0 \le i \le n$;

$\tilde{D}_n^2 = \{\tilde{d}_{ij}^{(2)}\}_{\substack{0 \le i \le n \\ 0 \le j \le n}}$ second derivative matrix in $\mathbf{P}_n$ with respect to the Gauss-Lobatto collocation nodes $\eta_i^{(n)}, \ 0 \le i \le n$;

Π_n orthogonal projection operator from $L^2(-1,1)$ into $\mathbf{P}_n$;

$\{\Theta_k^{(n)}\}_{0 \le k \le n^2 + 2n}$ Legendre Gauss-Lobatto nodes in $[-1,1] \times [-1,1]$;

$\tilde{I}_n$ interpolation operator at the collocation nodes;

$\mathbf{P}_n^\star$ space of polynomials in two variables of degree less or equal to n in each variable;

$\mathbf{P}_n^{\star,0}$ polynomials of $\mathbf{P}_n^\star$ vanishing on the boundary of the set $[-1,1] \times [-1,1]$;

$\{(\tau_{i,j}^{(n)}, \upsilon_{i,j}^{(n)})\}_{\substack{0 \le i \le n \\ 0 \le j \le n}}$ nodes of the upwind grid;

$\mathcal{L}_n$ discretization matrix of the differential operator with respect to the Legendre grid;

$\mathcal{A}_n$ discretization matrix of the differential operator including boundary conditions;

$\mathcal{B}_n$ finite-differences preconditioning matrix;

$\mathcal{S}_n$ shifting operator matrix from the Legendre grid to the upwind grid;

δ_t time step.

$\Gamma_k^{(m)}, \quad 1 \le k \le 4$ sides of the subdomain $\bar{\Omega}_m, \ 1 \le m \le M$;

$V_k^{(m)}, \quad 1 \le k \le 4$ vertices of the subdomain $\bar{\Omega}_m, \ 1 \le m \le M$;

References

ACHDOU Y. & KUZNETSOV Y. (1995), Substructuring preconditioners for finite element methods and nonmatching grids, East–West J. Numer. Meth., **3**, pp. 1–28.

ACHDOU Y. & PIRONNEAU O. (1996), A fast solver for Navier–Stokes equations in the laminar regime using mortar finite element and boundary element method, SIAM J. Numer. Anal., **32**, no. 4, pp. 985–1016.

AGOSHKOV V.I. (1988), Poincaré–Steklov's operators and domain decomposition methods in finite dimensional spaces, Proceedings of the First International Symposium on Domain Decomposition Methods for P.D.E.s (R.Glowinski, G.H.Golub, G.A.Meurant & J.Périaux Eds.), SIAM, Philadelphia.

AINSWORTH M. & ODEN J.T. (1992), A procedure for a posteriori error estimation for h–p finite element methods, Comput. Methods Appl. Mech. Engrg., **101**, pp. 73–96.

ALT H.W. (1980), Numerical solution of steady–state porous flow free boundary problems, Numer. Math., **36**, pp. 73–98.

ANAGNOSTOU G., MADAY Y. & PATERA A.T. (1997), A sliding mesh method for partial differential equations in nonstationary geometries, application to the Navier–Stokes equations, SIAM J. Numer. Anal., to appear.

BABUSKA I. & RHEINBOLDT W.C. (1978), A posteriori error estimates for the finite element method, Internat. J. Numer. Methods Engrg., **12**, pp. 1597–1615.

BAIOCCHI C., BREZZI F. & FRANCA L.P. (1993), Virtual bubbles and Ga.L.S., Comput. Methods Appl. Mech. Engrg., **105**, pp. 125–142.

BAIOCCHI C. & CAPELO A. (1984), *Variational and Quasivariational Inequalities*, John Wiley and Sons, New York.

BAIOCCHI C., COMINCIOLI V., GUERRI L. & VOLPI G. (1973), Free boundary problems in the theory of fluids through porous media: a numerical approach, Calcolo, Vol. X, fasc. 1, pp. 1–85.

BANK R.E. & SMITH R.K. (1993), A posteriori error estimates based on hierarchical bases, SIAM J. Numer. Anal., **30**, pp. 921–935.

BANK R.E. & WEISER A. (1985), Some a posteriori error estimates for elliptic partial differential equations, Math. Comp., **44**, pp. 238–301.

BATCHELOR G.K. (1967), *An Introduction to Fluid Dynamics*, Cambridge University Press, Cambridge.

BEAR J. (1962), *Dynamics of Fluids in Porous Media*, America Elsevier, New York.

BELHACHMI Z. & BERNARDI C. (1994), Resolution of fourth–order problems by the mortar element method, Comput. Methods Appl. Mech. Engrg., **116**, pp. 53–58.

BEN BELGACEM F. & MADAY Y. (1994), Non conforming spectral element methodology tuned to parallel implementation, Comput. Methods Appl. Mech. Engrg., **116**, pp. 59–67.

BERNARDI C. & MADAY Y. (1989), Approximation results for spectral methods with domain decomposition, Appl. Numer. Math., **6**, pp. 33–52.

BERNARDI C. & MADAY Y. (1992), *Approximations Spectrales de Problèmes aux Limites Elliptiques*, Mathématiques & Applications 10, Springer-Verlag, Paris.

BERNARDI C., MADAY Y. & PATERA A.T. (1997), Spectral Element Methods, in *Handbook of Numerical analysis* (P.G.Ciarlet & J.L.Lions Eds.), North Holland, Amsterdam, to appear.

BERNARDI C. & PELISSIER M.C. (1994), Spectral discretization of a mixed Schrödinger boundary–value problem and application, Finite Elements in Analysis and Design, **16**, pp. 309–315.

BERTOLUZZA S. & NALDI G. (1994), An adaptive wavelet collocation method, Report no. 932, I.A.N.–C.N.R., Pavia.

BIANCHI S. (1995), Approssimazione in 3–D con metodi spettrali, Thesis, University of Pavia, Italy.

BITSADZE A.V. (1968), *Boundary Value Problems for Second Order Elliptic Equations*, North Holland, Amsterdam.

BJØRSTAD P.E. & WIDLUND O.B. (1986), Iterative methods for the solution of elliptic problems on regions partitioned into substructures, SIAM J. Numer. Anal., **23**, pp. 1093–1120.

BLEANEY B.I. & BLEANEY B. (1976), *Electricity and Magnetism*, Third edition, Oxford University Press.

BLEISTEIN N. (1984), *Mathematical Methods for Wave Phenomena*, Computer Science and Applied Mathematics, Academic Press, Orlando.

BOFFI D. & FUNARO D. (1994), An alternative approach to the analysis and the approximation of the Navier–Stokes equations, J. Sci. Comput., **9**, no. 1, pp. 1–16.

BOYD J.P. (1989), *Chebyshev & Fourier Spectral Methods*, Lecture Notes in Engineering, Springer–Verlag, New York.

BRAMBLE J.H., PASCIAK J.E. & SCHATZ A.H. (1986), The construction of preconditioners for elliptic problems by substructuring, I, Math. Comp., **47**, pp. 103–134.

BRESSAN N.& QUARTERONI A. (1986), An implicit/explicit spectral method for Burgers equation, Calcolo, **23**, pp. 265–284.

BREZZI F. & GILARDI G. (1987), in *Finite Element Handbook* (H.Kardenstuncer Ed.), McGraw–Hill, New York, pp. 1–121.

BREZZI F. & MARINI L. D. (1992), A three field domain decomposition method, in *Domain Decomposition Methods in Science and Engineering* (A.Quarteroni, J.Periaux, Y.A.Kuznetsov & O.B.Widlund Eds.), AMS, Providence RI.

BREZZI F. & RUSSO A. (1994), Choosing bubbles for advection–diffusion problems, Math. Models Methods Appl. Sci., **4**, pp. 571–587.

BROOKS A.N. & HUGHES T.J.R. (1982), Streamline Upwind/ Petrov-Galerkin formulations for convection dominated flows with particular emphasis on the incompressible Navier–Stokes equations, Comput. Methods Appl. Mech. Engrg., **32**, pp. 199–259.

BRUCH J.C. JR & SLOSS J.M. (1989), Unsteady seepage through on earth dam with a toe drain, in *Finite Elements Analysis in Fluids* (T.J.Chung & G.R.Karr Eds.), UAH Press, Huntsville AL, pp. 1311–1316.

CANUTO C. (1988), Spectral methods and a maximum principle, Math. Comp., **51**, no. 184, pp. 615–629.

CANUTO C. (1994), Stabilization of spectral methods by finite element bubble functions, Comput. Methods Appl. Mech. Engrg., **116**, pp. 13–26.

CANUTO C. & FUNARO D. (1988), The Schwarz algorithm for spectral methods, SIAM J. Numer. Anal., **25**, no. 1, pp. 24–40.

CANUTO C., HUSSAINI M.Y., QUARTERONI A.& ZANG T.A. (1988), *Spectral Methods in Fluid Dynamics*, Springer Series in Computational Physics, Springer–Verlag, New York.

CANUTO C. & PUPPO G. (1994), Bubble stabilization of spectral Legendre methods for the advection–diffusion equations, Comput. Methods Appl. Mech. Engrg., **118**, pp. 239–263.

CANUTO C. & QUARTERONI A. (1982), Approximation results for orthogonal polynomials in Sobolev spaces, Math. Comp., **38**, pp. 67–86.

CANUTO C. & QUARTERONI A. (1985), Preconditioned minimal residual methods for Chebyshev spectral calculations, J. Comput. Phys., **60**, pp. 315–337.

CANUTO C. & VAN KEMENADE V. (1996), Bubble–stabilized spectral methods for the incompressible Navier–Stokes equations, Comput. Methods Appl. Mech. Engrg., **135**, pp. 35–61.

CHAN T.F. & GOOVAERTS D. (1989), Shur complement domain decomposition algorithms for spectral methods, Appl. Numer. Math., **6**, pp. 53–64.

CHAN T.F., MATHEW T.P. & SHAO J.P. (1994), Efficient variants of the vertex space domain decomposition algorithm, SIAM J. Sci. Comput., **15**, no. 6, pp. 1349–1374.

COURANT R. & HILBERT D. (1953), *Methods of Mathematical Physics*, Wiley–Interscience, New York.

DAUTRAY R. & LIONS J.L. (1988–1993), *Mathematical Analysis and Numerical Methods for Science and Technology*, 6 Volumes, Springer–Verlag, Heidelberg.

DELFOUR M., FORTIN M. & PAYRE G. (1981), Finite–difference solutions of a non–linear Schrödinger equation, J. Comput. Phys., **44**, pp. 277–288.

DEMARET P., DEVILLE M.O. & SCHNEIDESCH C. (1989), Thermal convection solution by Chebyshev pseudospectral multi–domain decomposition and finite element preconditioning, Appl. Numer. Math., **6**, pp. 107–121.

DENNEMEYER R. (1968), *Introduction to Partial Differential Equations and Boundary Value Problems*, Mc Graw–Hill, New York.

DE VERONICO M.C., FUNARO D. & REALI G.C. (1994), A novel numerical technique to investigate nonlinear guided waves: approximation of nonlinear Schrödinger equation by nonperiodic pseudospectral methods, Numer. Methods Partial Differential Equations, **10**, no. 6, pp. 667–675.

DEVILLE M. & MUND E. (1991), Finite element preconditioning of collocation schemes for advection–diffusion equations, Proc. IMACS Int. Symp. Iterative Methods in Linear Algebra (R.Beauwens & P. de Groen Eds.), North Holland, Amsterdam, pp. 181–190.

DEVILLE M. & MUND E. (1992), Fourier analysis of finite element preconditioned collocation schemes, SIAM J. Sci. Statist. Comput., **13**, p.596.

DOUGLAS J. JR & WANG J. (1989), An absolutely stabilized finite element method for the Stokes problem, Math. Comp., **52**, pp. 495–508.

DRYJA M. (1984), A finite–element capacitance method for elliptic problems on regions partitioned into subregions, Numer. Math., **44**, pp. 153–168.

DRYJA M., SMITH B.F. & WIDLUND O.B. (1994), Schwarz analysis of iterative substructuring algorithms for elliptic problems in three dimensions, SIAM J. Numer. Anal., **31**, no. 6, pp. 1662–1694.

DU FORT E.C. & FRANKEL S.P. (1953), Stability conditions in the numerical treatment of parabolic differential equations, Math. Tables and other Aids to Computation, **7**, p.135.

EHRENSTEIN U., GUILLARD H. & PEYRET R. (1989), Flame computations by a Chebyshev multidomain method, Internat. J. Numer. Methods Fluids, **9**, pp. 499–515.

EHRENSTEIN U. & PEYRET R. (1989), A Chebyshev collocation method for the Navier–Stokes equations with application to double–diffusive convection, Internat. J. Numer. Methods Fluids, **9**, pp. 427–452.

FARCY A. & ALZIARY DE ROQUEFORT T. (1990), Pseudospectral multidomain method for incompressible viscous flow computation, in *Spectral and High Order Methods for Partial Differential Equations* (C.Canuto & A.Quarteroni Eds.), North Holland, Amsterdam, pp. 337–346.

FISCHER P.F. & RØNQUIST E.M. (1994), Spectral element methods for large scale parallel Navier–Stokes calculations, Comput. Methods Appl. Mech. Engrg., **116**, pp. 69–76.

FORNBERG B. (1996), *A Practical Guide to Pseudospectral Methods*, Cambridge Monographs on applied and Computational Mathematics, v. 1, Cambridge University Press.

FORNBERG B. & SLOAN D.M. (1994), A review of pseudospectral methods for solving partial differential equations, Acta Numerica, pp. 203–267.

FUNARO D. (1986), A multidomain spectral approximation of elliptic equations, Numer. Methods Partial Differential Equations, **2**, pp. 187–205.

FUNARO D. (1987), A preconditioning matrix for the Chebyshev differencing operator, SIAM J. Numer. Anal., **24**, pp. 1024–1031.

FUNARO D. (1988), Domain decomposition methods for pseudospectral approximations. Part I: second order equations in one dimension, Numer. Math., **52**, pp. 329–344.

FUNARO D. (1992), *Polynomial Approximation of Differential Equations*, Lecture Notes in Physics, m8, Springer–Verlag, Heidelberg.

FUNARO D. (1993), A new scheme for the approximation of advection–diffusion equations by collocation, SIAM J. Numer. Anal., **30**, no. 6, pp. 1664–1676.

FUNARO D. (1993b), FORTRAN routines for spectral methods, Report no. 891, I.A.N.–C.N.R., Pavia. Available with software on anonymous ftp server: ian.pv.cnr.it, in the directory /pub/splib.

FUNARO D. (1997), Some remarks about the collocation method on a modified Legendre grid, An Int. J. on Computers and Math. with Applications, to appear.

FUNARO D., GIANGI M. & MANSUTTI D. (1997), A splitting method for unsteady incompressible viscous fluids imposing no boundary conditions to pressure, J. Sci. Comput., to appear.

FUNARO D. & GOTTLIEB D. (1988), A new method of imposing boundary conditions in pseudospectral approximations of hyperbolic equations, Math. Comp., **51**, no. 184, pp. 599–613.

FUNARO D., QUARTERONI A. & ZANOLLI P. (1988), An iterative procedure with interface relaxation for domain decomposition methods, SIAM J. Numer. Anal., **25**, pp. 1213–1236.

FUNARO D. & ROTHMAN E. (1989), Preconditioning matrices for the pseudo spectral approximation of first–order operators, in *Finite Elements Analysis in Fluids* (T.J.Chung & G.R.Karr Eds.), UAH Press, Huntsville AL, pp. 1458–1463.

FUNARO D. & RUSSO A. (1993), Approximation of advection–diffusion problems by a modified Legendre grid, in *Finite Elements in Fluids, New Trends and Applications* (K.Morgan, E.Oñate, J.Periaux, J.Peraire & O.C.Zienkiewicz Eds.), Pineridge Press, pp. 1311–1318.

GARBA A., MOFID A. & PEYRET R. (1994), Spectral solution of free surface flows, Comput. Methods Appl. Mech. Engrg., **116**, pp. 331–346.

GOTTLIEB D., HUSSAINI M.Y. & ORSZAG S.A. (1984), Theory and application of spectral methods, in *Spectral Methods for Partial Differential Equations* (R.G.Voigt, D.Gottlieb & M.Y.Hussaini Eds.), SIAM–CBMS, Philadelphia, pp. 1–54.

GOTTLIEB D. & ORSZAG S.A. (1977), *Numerical Analysis of Spectral Methods, Theory and Applications*, CBMS–NSF Regional Conference Series in Applied Mathematics, SIAM, Philadelphia.

GUMMEL H.K. (1964), A self–consistent iterative scheme for one–dimensional steady state transistor calculations, IEEE Trans. on Electron Devices, *ED-11*, pp. 455–465.

HAAR M.E. (1962), *Groundwater ans Seepage*, Mc Graw–Hill, New York.

HAIDVOGEL D.B. & ZANG T.A. (1979), An accurate solution of Poisson's equation by expansion in Chebyshev polynomials, J. Comput. Phys., **30**, pp. 167–180.

HALDENWANG P. (1984), Résolution tridimensionnelle des équations de Navier-Stokes par méthodes spectrales Tchébycheff: application á la convection naturelle, Ph.D Thesis, Université de Provence.

HANLEY P. (1993), A strategy for efficient simulation of viscous compressible flows using a multi-domain pseudo-spectral method, J. Comput. Phys., **108**, no. 1, pp. 153–158.

HASEGAWA A. (1989), *Optical Solitons in Fibers*, Second enlarged edition, Springer-Verlag.

HO L.W. & RØNQUIST E.M. (1994), Spectral element solution of steady incompressible free-surface flows, Comput. Methods Appl. Mech. Engrg., **116**, pp. 347–367.

HUGHES T.J.R., FRANCA L.P. & HULBERT G.M. (1989), A new finite element formulation for computational fluid dynamics: VIII. The Galerkin/Least-Square method for advective-diffusive equations, Comput. Methods Appl. Mech. Engrg., **73**, pp. 173–189.

ISRAELI M., VOZOVOI L. & AVERBUCH A. (1993), Domain decomposition methods for solving parabolic PDEs on multiprocessors, Appl. Numer. Math., **12**, pp. 193–212.

JAIN M.K. (1984), *Numerical Solution of Differential Equations*, Second edition, John Wiley & Sons, New York.

JOHNSON C. (1987), *Numerical Solutions of Partial Differential Equations by the Finite Element Method*, Cambridge University Press, Cambridge.

JOHNSON C. (1990), Adaptive finite element methods for diffusion and convection problems, Comput. Methods Appl. Mech. Engrg., **82**, pp. 301–322.

KARNIADAKIS G.E. (1989), Spectral element simulations of laminar and turbulent flows in complex geometries, Appl. Numer. Math., **6**, pp. 85–105.

KOPRIVA D.A. (1986), A spectral multidomain method for the solution of hyperbolic systems, Appl. Numer. Math., **2**, pp. 221–241.

KOPRIVA D.A. (1991), Multidomain spectral solution of the Euler gasdynamics equations, J. Comput. Phys., **96**, no. 2, pp. 428–450.

KORCZAK K.Z. & PATERA A.T. (1986), Isoparametric spectral element method for solution of the Navier-Stokes equations in complex geometry, J. Comput. Phys., **62**, pp. 361–382.

KREISS H.O. & LORENZ J. (1989), *Initial-Boundary Value Problems and the Navier-Stokes Equation*, Academic Press, London.

KU H.C., HIRSH R.S., TAYLOR T.D. & ROSEMBERG A.P. (1989), A pseudospectral matrix element method for solution of three–dimensional incompressible flows and its parallel implementation, J. Comput. Phys., **83**, pp. 260–291.

LACROIX J.M., PEYRET R. & PULICANI J.P. (1988), A pseudospectral multidomain method for the Navier–Stokes equations with applications to double diffusive convection, Proc. 7th GAMM Conf. Numer. Meth. Fluid Mech. (M.Deville Editor), Vieweg–Verlag, Braunschweig, pp. 167–174.

LADYZHENSKAYA O.A. (1969), *The Mathematical Theory of Viscous Incompressible Flow* (translated from russian), Second edition, Gordon and Breach Science Publishers, New York.

LANDAU L.D. & LIFSCHITZ E.M. (1959), *Course of Theoretical Physics, Volume 6, Fluid Mechanichs*, Pergamon Press, Oxford.

LIONS J.L. & MAGENES E. (1972), *Non–Homogeneous Boundary Value Problems and Applications*, Springer–Verlag, New York.

MACARAEG M.C. & STREETT C.L. (1986), Improvements in spectral collocation through a multiple domain technique, Appl. Numer. Math., **2**, pp. 95–108.

MADAY Y. (1989), Relèvement de traces plynômiales et interpolations hilbertiennes entre espaces de polynômes, C.R. Acad. Sci. Paris, t. 309, Série I, pp. 463–468.

MADAY Y. (1991), Résultats d'approximation optimaux pur les opérateurs d'interpolation polynomiale, C.R. Acad. Sci. Paris, t. 312, Série I, pp. 705–710.

MADAY Y., MAVRIPLIS C.A. & PATERA A.T. (1989), Non conforming mortar element method: application to spectral discretizations, in *Domain Decomposition Methods* (T.F.Chan, R.Glowinski, J.Periaux & O.B.Widlund Eds.), SIAM, Philadelphia.

MADAY Y., PERRIER V. & RAVEL J.C. (1991), Adaptivité dynamique sur bases d'ondelettes pour l'approximation d'équations aux derivées partielles, C.R. Acad. Sci. Paris, t. 312, Série I, pp. 405–410.

MADAY Y. & RØNQUIST E.M. (1990), Optimal error analysis of spectral methods with emphasis on non–constant coefficients and deformed geometries, in *Spectral and High Order Methods for Partial Differential Equations* (C.Canuto & A.Quarteroni Eds.), North Holland, Amsterdam, pp. 91–115.

MANDEL J. (1994), An iterative solver for *p*–version finite elements in three dimensions, Comput. Methods Appl. Mech. Engrg., **116**, pp. 175–183.

MANTEUFFEL T.A. (1977), The Chebyshev iteration for nonsymmetric linear systems, Numer. Math., **28**, pp. 307–327.

MANTEUFFEL T.A. (1978), Adaptive procedure for estimating parameters for nonsymmetric Chebyshev iteration, Numer. Math., **31**, pp. 183–208.

MARINI L.D. & PIETRA P. (1986), Fixed–point algorithm for stationary flow in porous media, Comput. Methods Appl. Mech. Engrg., **56**, pp. 17–45.

MARKOWICH P.A. (1986), *The Stationary Semiconductor Device Equations*, Springer, Wien–New York.

MAVRIPLIS C. (1989), Nonconforming discretization and a posteriori estimations for adaptive spectral element techniques, Ph.D. Thesis, M.I.T., Cambridge MA.

METIVET B. (1989), A curved multidomain spectral collocation method for solving Navier–Stokes equations, in *Finite Elements Analysis in Fluids* (T.J.Chung & G.R.Karr Eds.), UAH Press, Huntsville AL, pp. 1422–1427.

METIVET B. & MORCHOISNE Y. (1982), Multi–domain spectral technique for viscous flow calculations, Proc. 4th GAMM Conf. Numer. Methods Fluid Mech. (M.Deville Editor), Viewig–Verlag, Braunschweig.

MEYER R.E. (1971), *Introduction to Mathematical Fluid Dynamics*, Dover Publications, New York.

MOFID A. & PEYRET R. (1992), Stability of the collocation–Chebyshev approximation to the advection–diffusion equation, Report no. 320, Université de Nice.

MONTIGNY–RANNOU F. & MORCHOISNE Y. (1987), A spectral method with staggered grid for incompressible Navier–Stokes equations, Internat. J. Numer. Methods Fluids, **7**, pp. 175–189.

MORCHOISNE Y. (1983), Résolution des équations de Navier Stokes par une méthode spectrale de sous–domaines, Proc. III Internat. Congress Numer. Methods Engrg. (P.Lascaux Editor), Paris.

NATARAJAN R. (1995), Domain decomposition using spectral expansions of Steklov–Poincaré operators, SIAM J. Sci. Comput., **16**, no. 2, pp. 470–495.

ORSZAG S.A. (1980), Spectral methods for problems in complex geometries, J. Comput. Phys., **37**, pp. 70–92.

PASQUARELLI F. & QUARTERONI A. (1994), Effective spectral approximations of convection–diffusion equations, Comput. Methods Appl. Mech. Engrg., **116**, pp. 39–51.

PATERA A.T. (1984), A spectral element method for fluid dynamics: laminar flow in a channel expansion, J. Comput. Phys., **54**, pp. 468–488.

PAVARINO L.F. & WIDLUND O.B. (1996), A polylogarithmic bound for an iterative substructuring method for spectral elements, SIAM J. Numer. Anal., **33**, pp. 1303–1335.

PEYRET R. (1990), The Chebyshev multidomain approach to stiff problems in fluid mechanics, Comput. Methods Appl. Mech. Engrg., **80**, pp. 129–145.

PEYRET R. & TAYLOR T.D. (1982), *Computational Methods for Fluid Flow*, Springer–Verlag.

PHILLIPS T.N. & KARAGEORGHIS A. (1989), Efficient direct methods for solving the spectral collocation equations for Stokes flow in rectangularly decomposable domains, SIAM J. Sci. Statist. Comput., **10**, pp. 89–103.

PINELLI A., BENOCCI C. & DEVILLE M. (1994), Chebyshev pseudo–spectral solution of advection–diffusion equations with mapped finite difference preconditioning, J. Comput. Phys., **112**, no. 1, pp. 1–11.

PINELLI A., COUZY W., DEVILLE M.O. & BENOCCI C. (1996), An efficient iterative solution method for the Chebyshev collocation of advection–dominated transport problems, SIAM J. Sci. Comput., **17**, no. 3, pp. 647–657.

PROSKUROWSKI W. & WIDLUND O. (1976), On the numerical solution of Helmholtz's equation by the capacitance matrix method, Math. Comp., **30**, no. 135, pp. 433–468.

PULICANI J.P. (1988), A spectral multi–domain method for the solution of the 1–D Helmholtz and Stokes–type equations, Comput. & Fluids, **16**, pp. 207–215.

QUARTERONI A. & SACCHI-LANDRIANI G. (1988), Domain decomposition preconditioners for the spectral collocation method, J. Sci. Comput., **3**, no. 1, pp. 45–76.

QUARTERONI A. & VALLI A. (1994), *Numerical Approximation of Partial Differential Equations*, Springer Series in Computational Math., v. 23, Springer–Verlag, Heidelberg.

RICHTMYER R.D. & MORTON K.W. (1967), *Difference Methods for Initial-Value Problems*, Interscience, John Wiley & Sons, New York.

RØNQUIST E.M. (1988), Optimal spectral element methods for the unsteady three–dimensional incompressible Navier–Stokes equations, Ph.D. Thesis, M.I.T., Cambridge MA.

RØNQUIST E.M. (1996), Convection treatment using spectral elements of different order, Internat. J. Numer. Methods Fluids, **22**, pp. 241–264.

RUSSO A. (1996), A posteriori error estimators via bubble functions, Math. Models Methods Appl. Sci., **6**, pp. 33–41.

SCHNEIDESCH C.R. & DEVILLE M.O. (1994), Multidomain decomposition of curved geometries in the Chebyshev collocation method for thermal problems, Comput. Methods Appl. Mech. Engrg., **116**, pp. 87–94.

SELBERHERR S. (1984), *Analysis and Simulation of Semiconductors Devices*, Springer, Wien–New York.

SMITH B.F. (1992), An optimal domain decomposition preconditioner for the finite element solution of linear elasticity problems, SIAM J. Sci. Statist. Comput., **13**, pp. 364–378.

SMITH R.A. (1978), *Semiconductors*, Second Edition, Cmbridge University Press, Cambridge.

SMOLLER. J. (1983), *Shock Waves and Reaction–Diffusion Equations*, Springer-Verlag, New York.

SOBOLEV S.L. (1964), *Partial Differential Equations of Mathematical Physics*, Pergamon Press, Oxford.

STRAUSS W.A. (1978), The nonlinear Schrödinger equation, in *Contemporary Developments in Continuum Mechanics and Partial Differential Equations* (G.M. de La Penha, L.A. Medeiros Eds.), North Holland, Amsterdam.

SULEM P.L., SULEM C. & PATERA A.T. (1984), Numerical simulation of singular solutions to the two–dimensional cubic Schrödinger equation, Comm. Pure Appl. Math., **37**, pp. 755–778.

SZE S.M. (1981), *Physics of Semiconductor Devices*, Second Edition, John Wiley & Sons, New York.

TEMAM R. (1985), *Navier–Stokes equations*, Second edition, North–Holland, Amsterdam.

VERFÜRTH R. (1994), A posteriori error estimation and adaptive mesh–refinement techniques, J. Comput. Appl. Math., **50**, pp. 67–83.

WEINBERGER H.F. (1965), *A first Course in Partial Differential Equations*, John Wiley & Sons, New York.

WEINSTEIN M.I. (1983), Nonlinear Schrödinger equations and sharp interpolation estimates, Comm. Math. Phys., **87**, pp. 567–576.

WHITHAM G.B. (1974), *Linear and Nonlinear Waves*, Interscience, John Wiley & Sons, New York.

XU J. (1992), Iterative methods by space decomposition and subspace correction, SIAM Rev., **34**, pp. 581–613.

YUEN H.C. & LAKE B.M. (1978), Nonlinear wave concepts applied to deep-water waves, in *Solitons in Action* (K. Lonngren & A. Scott Eds.), Academic Press, New York.

ZANG T.A. & HUSSAINI M.Y. (1985), Recent application of spectral methods in fluid dynamics, in *Large Scale Computations in Fluid Mechanics*, Lectures in Appl. Math., **22**, pp. 379–409.

ZAUDERER E. (1983), *Partial Differential Equations of Applied Mathematics*, Interscience, John Wiley & Sons, New York.

Index

Springer
und
Umwelt

Als internationaler wissenschaftlicher
Verlag sind wir uns unserer besonderen
Verpflichtung der Umwelt gegenüber
bewußt und beziehen umweltorientierte
Grundsätze in Unternehmens-
entscheidungen mit ein. Von unseren
Geschäftspartnern (Druckereien,
Papierfabriken, Verpackungsherstellern
usw.) verlangen wir, daß sie sowohl
beim Herstellungsprozess selbst als
auch beim Einsatz der zur Verwendung
kommenden Materialien ökologische
Gesichtspunkte berücksichtigen.
Das für dieses Buch verwendete Papier
ist aus chlorfrei bzw. chlorarm
hergestelltem Zellstoff gefertigt und im
pH-Wert neutral.

 Springer

Editorial Policy

§1. Submissions are invited in the following categories:

i)   Research monographs
ii) Lecture and seminar notes
iii) Reports of meetings

Those considering a project which might be suitable for the series are strongly advised to contact the publisher or the series editors at an early stage.

§2. Categories i) and ii). These categories will be emphasized by Lecture Notes in Computational Science and Engineering. Submissions by interdisciplinary teams of authors are encouraged. The goal is to report new developments – quickly, informally, and in a way that will make them accessible to non-specialists. In the evaluation of submissions timeliness of the work is an important criterion. Texts should be well-rounded and reasonably self-contained. In most cases the work will contain results of others as well as those of the authors. In each case the author(s) should provide sufficient motivation, examples, and applications. In this respect, articles intended for a journal and Ph.D. theses will usually be deemed unsuitable for the Lecture Notes series. Proposals for volumes in this category should be submitted either to one of the series editors or to Springer-Verlag, Heidelberg, and will be refereed. A provisional judgment on the acceptability of a project can be based on partial information about the work: a detailed outline describing the contents of each chapter, the estimated length, a bibliography, and one or two sample chapters – or a first draft. A final decision whether to accept will rest on an evaluation of the completed work which should include

- at least 100 pages of text;
- a table of contents;
- an informative introduction perhaps with some historical remarks which should be accessible to readers unfamiliar with the topic treated;
- a subject index.

§3. Category iii). Reports of meetings will be considered for publication provided that they are both of exceptional interest and devoted to a single topic. In exceptional cases some other multi-authored volumes may be considered in this category. One (or more) expert participants will act as the scientific editor(s) of the volume. They select the papers which are suitable for inclusion and have them individually refereed as for a journal. Papers not closely related to the central topic are to be excluded. Organizers should contact Lecture Notes in Computational Science and Engineering at the planning stage.

§4. Format. Only works in English are considered. They should be submitted in camera-ready form according to Springer-Verlag's specifications. Electronic material can be included if appropriate. Please contact the publisher. Technical instructions and/or TeX macros are avaiable on http://www.springer.de/author/tex/help-tex.html; the name of the macro package is "LNCSE – LaTeX2e class for Lecture Notes in Computational Science and Engineering". The macros can also be sent on request.